Regional Security in Southeast Asia and the South Pacific

Prospects of Nuclear Free Zone

Regional Security in Southeast Asia and the South Pacific

Prospects of Nuclear Free Zone

Amba Pande

AUTHORSPRESS

First Published 2002
by Authorspress

E-35/103, Jawahar Park
Laxmi Nagar, Delhi-110 092

e-mail: authorspress@yahoo.com
quantum_soul@mantramail.com

Typeset in NewCenturySchlbk by Chitra Computers

Regional Security in Southeast Asia and the South Pacific: Prospects of Nuclear Free Zone
ISBN 81-7273-101-9

Printed in India at Tarun Offset Press, Delhi.

Dedicated to

Bade Babuji—Badi Amma
and
Babuji—Amma

Preface

Ever since the bombing of Hiroshima and Nagasaki in 1945 mankind has become unforgettably aware of the existence of nuclear weapons as a means of mass annihilation. Nuclear weapons, their testing, development, deployment and threat of use have posed the biggest challenge to the human existence. This realization has also highlighted the urgent need for comprehensive nuclear disarmament and banning the use and production of these weapons. Unfortunately this goal has not been achieved within last fifty six years. On the contrary, nuclear arsenals have continued to grow in numbers and their destructive capacity and have become the most serious threat to human security.

The breakdown of bi-polar world and related tensions have hardly brought any relief to the nuclear dangers faced by the world. We are still faced with the fear of horizontal as well as vertical proliferation (as India and Pakistan have entered the nuclear club and the United States has proposed its National Missile Defence Programme). Therefore, it is imperative on moral, political and legal grounds and also for reasons of security to abolish these weapons. In fact, it is one of the most serious and urgent tasks before us and any progress in this direction demands obligations and responsibility on part of both nuclear and non-nuclear weapon states at all levels and in every region.

In this context, regional approaches to nuclear arms control such as nuclear free zones are significant. These zones ban the manufacture and deployment of nuclear weapons in specific regions and seek guarantee from the nuclear armed states that the zones would not be threatened or attacked by nuclear weapons. Nuclear free zones can be considered as an important first step towards the total and complete elimination of nuclear weapons. Moreover, the creation of such zones also contributes to confidence building and regional cooperation. In view of this, the present study will endeavour to discuss the evolution of the concept of nuclear free zone and to the extent it can play a role in the regional security framework of the region concerned. Against this backdrop, the cases of Southeast Asia and South Pacific Nuclear Free zones would be worth discussing. These regions were exposed to nuclear dangers directly or indirectly right since the beginning of the nuclear era.

In the case of Southeast Asia due to excessive external intervention, the cold war era seeped into the region in such a way that the regional conflicts almost became an extension of the Super Power rivalry. Be it the Vietnam war or the Cambodian imbroglio, the situation flexed up because of the intervention of the great powers. The situation also polarised the regional countries but at the same time it also made them realise the need for creating a regional order to regulate the conduct of external as well as regional countries. The Zone of Peace Freedom and Neutrality (ZOPFAN), proposal was an endeavour by the ASEAN countries to achieve the above mentioned objectives. The nuclear weapon free zone for Southeast Asia developed as a component of ZOPFAN since the nuclear danger was implicit in the overall threat environment.

In the South Pacific, unlike Southeast Asia there was an absence of conventional military threat but the nuclear danger existed in the form of threat to environmental and human security. The region got exposed to nuclear danger since the beginning of nuclear age. The nuclear tests conducted by the US, Britain and France made the people suffer physically as well environmentally. There was also a fear of the South Pacific becoming a dustbin for nuclear wastes. Besides, the presence of the US bases with nuclear capabilities made Australia a probable target for nuclear attack. It is in this background that the idea of Nuclear-free zone was evolved as an effort by the regional countries to protect themselves from the nuclear threat.

In the present study the Nuclear Free Zones have been discussed in the context of regional security in Southeast Asia and the South Pacific. It is divided into four parts. Part I deals with the conceptual framework of nuclear weapon free zones. Part II gives a detailed explanation of the regional security order in Southeast Asia and the factors leading to the proposal of ZOPFAN and SEANWFZ, their evolution and prospects. In Part III the South Pacific Nuclear Free Zone has been discussed along with regional security order before and after the treaty came into existence. Part IV is the conclusion and highlights the relevance of NWFZ concept in general and for Southeast Asia and the South Pacific in particular.

I am indebted to several people for their help and advice during the course of this study. First of all I express my sincere gratitude towards my guide Dr. Ganganath Jha for supervising this work. I am also thankful to Dr. Man Mohini Kaul for encouraging me to get this book published. My sincere thanks are also due to Dr Ramesh Thakur and Prof. Ian Talbot who were kind enough to grant me personal interviews and enlighten me on the subject. I am also grateful to my brother Sanjay, Naveen Bhaiya, Mr. Om Prakash, Man Singh and all my friends and well wishers. But above all I would like to thank my son Aarsh and daughter Ishita for bearing with me during my long working hours and my husband Shailendra who has helped me in every possible manner at every stage of the preparation of this book.

Amba Pande

Contents

PART I

Conceptual Framework of Nuclear Weapon-free Zones

1

Nuclear Weapon-free Zone: A Historical Background

The dawn of nuclear age brought with it the realisation of the tremendous potential of nuclear power for the betterment of mankind and simultaneously its capacity for global annihilation.[1] The nuclear bombing of Hiroshima and Nagasaki added a new chapter in international politics—the emergence of two power blocs. The two blocs engaged themselves in an ideological conflict where the military, political, economic, diplomatic, cultural and scientific forces of the two sides were poised against each other in a state of total confrontation. Thus, by 1947 the Cold War era began to set in, which created a 'state of insecurity' and gave rise to an arms race which included conventional as well as nuclear arms. Once started, this arms race reached a stage where nations began to believe that sophisticated nuclear arms are a deterrent, and peace can only be maintained through a balance of nuclear terror.

The nations believed that deterrence has served, and continues to serve, as an essential component in maintaining overall equilibrium between the forces of two major military alliances and has thereby contributed to strategic stability. This resulted in a continuous drive to produce varied, accurate and effective nuclear weapon systems. A step towards the further sophistication of nuclear arsenal on one side, however, was followed by similar efforts on the other giving rise to the so-called 'Vertical Proliferation'.

Under these circumstances nuclear weapons became an international currency of power. Monitored by the 'manipulative force' of the two Superpowers, the rest of the globe started channellising its potential resources towards a malevolent arms race further contributing to the fear of 'horizontal proliferation'. In 1945 only the United States had the technology to produce nuclear weapons but within a few years the Soviet Union, developed nuclear capability (1949) followed later by the United Kingdom (1952), France (1960), and China (1964). The original atomic or fission bomb was followed by the more potent hydrogen or thermonuclear fusion bomb.

In 1998 India and Pakistan jolted the world by conducting nuclear tests and declaring themselves as nuclear powers. There is also a growing concern over nuclear ambitions of Iraq, Iran and North Korea. These concerns have stimulated (or perhaps have become a justification for) a new enthusiasm for 'vertical proliferation'. The most significant move in this direction is the United States' effort towards setting up The National Missile Defence (NMD) and Theatre Missile Defence (TMD) Systems. This may create suspicions, uncertainties and could probably ignite a new arms race. Thus, after a decade of the end of the Cold War, there appears to be a real and growing danger that the world could drift into a second nuclear age just when there are compelling reasons to eradicate the scourge of nuclear weapons.[2]

Against this background, the urgency and relevance of nuclear disarmament and complete banishment of all nuclear weapons from the face of the earth becomes a matter of prime importance. It is, therefore, a responsibility of each and every country to prevent the worldwide proliferation. However, any progress in this direction, demands obligations and responsibilities on the part of both nuclear weapon and non-nuclear weapon states.

It may be mentioned, in this regard, that there had been several efforts in this direction. In fact, proposals for measures concerning the cessation of nuclear arms race go back to the early period of armament efforts. Over the years, a very wide spectrum of measures have been proposed in the United Nations and other multilateral fora covering limitations, reductions and the elimination of nuclear weapons and their delivery systems. Several bilateral as well as multilateral agreements have been signed among different countries such as the Hotline Agreement (1963), Anti Ballistic Missile Treaty (ABM) (1972), and Strategic Arms Limitation Talks I & II (SALT I & II) which are bilateral agreements, whereas, the multilateral agreements include the Antarctic Treaty (1959), the Partial Test Ban Treaty (PTBT) (1963), Outer Space Treaty (1967), the Non-Proliferation Treaty (NPT) (1968) and Seabed Treaty (1971).[3]

Besides, the above-mentioned bilateral and multilateral treaties, there has been tremendous outpouring of new and revived ideas for curbing the nuclear arms race. One of such approaches has been the regional approach such as the establishment of zones of peace and nuclear weapon-free zones or nuclear free zones. The objective of this kind of approach has been to ensure a certain level of security and stability in otherwise 'hot zones' of conflict. The nuclear weapon-free zones intend to ban the manufacture, deployment and transit or stationing of nuclear weapons in the specific regions, and to demand from the nuclear weapon states that such zones should not be threatened or attacked by nuclear weapons. The creation of nuclear free zones not only limits proliferation but also support active nuclear disarmament with the dismantling of nuclear weapons and fissile stocks and rolling back existing nuclear programme.[4] As we are aware, such zones already exist in Latin America, the South Pacific, Southeast Asia and Africa, and proposals exist for several other regions such as Central Asia, South Asia, Central Europe, and Northern Europe. Although the growth of these zones has not eliminated all conflicts or achieved general and complete disarmament in their areas, they have accomplished much in laying the foundation for the proliferation of peace and have asserted the rights of humanity to live in a nuclear weapon-free world.[5]

The idea of establishing nuclear free zone began to attract the attention of the international community during the 1950s. However, the early initiatives were limited to the search of formulae to prevent the proliferation of nuclear weapons and their genesis can be traced to the Soviet proposal submitted to the Disarmament Commission in 1956 concerning Central Europe. It was a zonal approach to the question of limiting nuclear armaments.[6] However, the first detailed proposal, in this regard, again for Central Europe, came up in 1956 which was contained in a plan submitted by the Polish Minister of Foreign Affairs, Adam Repacki.[7] The Repacki Plan aimed at 'elimination of nuclear weapons nuclear delivery vehicles, a reduction of military forces and conventional armaments.' The plan took within its purview the territories of Poland, former Czechoslovakia, and border areas of both the Germanys and it was to be carried out in two stages. In the first stage all nuclear armaments were to be freezed whereas, in the second, all nuclear and conventional armaments were to be eliminated. The Repacki Plan was an important step towards achieving the goal of disarmament and promoting security in the tension-prone areas of Europe. It was generally acceptable to smaller and neutral states of Europe. The United States, however, rejected this plan. The former Soviet Union, on the other hand, agreed to support it provided the United States, the United Kingdom, and France did the same. Some other proposals which came forward were the Soviet Bulgarian proposal of 1958 for Nordic nuclear weapon-free zone and Soviet proposal of 1963 for Mediterranean nuclear free zone. However, another concrete attempt, in this direction, was made by Finland. Finland's President Kekkonen suggested the creation of a nuclear weapon-free zone to cover Denmark, Finland, Iceland, Norway and Sweden. But, due to the difference in opinions and cold war politics, the effort could not attain its goal.[8]

With the escalation of Superpower rivalry outside Europe, the proposals for establishing nuclear free zones in other regions also started coming in. In 1956, a proposal was put forward to denuclearise Africa. In the Baltic, the demand to establish nuclear free zone dates back to the early fifties. In the Middle East, the proposal was put forward by Egypt and Iran in 1974. From South Asia, Pakistan made a proposal in 1974. Apart from these proposals, several treaties were also signed declaring particular regions as nuclear free zones and these are: the Antarctic Treaty (1959), the Treaty on Principles Governing the Activities of States in the Exploration and Use of Outer Space Including Moon and other Celestial Bodies (1967), the Treaty of Tlatelolco (1967), the Treaty of Prohibition of the Emplacement of Nuclear Weapons on Sea-bed and Ocean Floor and the Subsoil thereof (1971), the Rarotonga Treaty (1985), the Bangkok Treaty (1995), and the Treaty of Pelindaba (1996).

The question of nuclear free zone was also taken up by the Non-Aligned Movement (NAM) as early as 1972 in the George Town Summit.[9] The efforts of the countries of NAM were aimed at procuring a system for the regionalisation of peace and security based on the principles of non-interference of world powers in the regional or local conflict as well as the cessation of their indirect confrontation in interstate and intra-regional hostilities.[10] The United Nations also took up the question and the General Assembly passed several resolutions with regard to the establishment of nuclear free zone in different areas of the world. However, it was only during the early seventies that the General Assembly decided to conduct a detailed study on this subject. In 1974, the United Nations General Assembly, on the initiative taken by Finland, called for a 'comprehensive study of nuclear weapon-free zone in all its aspects' by an ad hoc group of governmental experts under the auspices of the Conference of the Committee on Disarmament (CCD).[11] The ad hoc group held a series of meetings and subsequently adopted the final study in its 25th meeting held on 8 April 1975.

The General Assembly reconsidered the study at its 30th Session in 1975 and passed it as Resolution 3472 A (XXX) on 11 December 1975, and recommended it for the attention of all governments, IAEA and other concerned international organisations.[12] After much discussion, the definition of nuclear weapons-free zone was adopted on the initiative of Mexico.

REFERENCES

1. See "The Perpetual Menace to Human Society", *Comprehensive Study on Nuclear Weapons Disarmament Study Series I* (New York: United Nations, 1981), p.145.
2. *"International Seminar on Nuclear Weapon Free Zones"*, A Concept paper published in *Development and Dialogue*, p.37.
3. For details of these treaties *see, The United Nations and Disarmament*, 1945-1985 (New York: United Nations, 1985).
4. "The Uppsala Declaration on Nuclear Weapon-Free Zones", published in *Development and Dialogue* 2000, 1-2, pp.46-48.
5. Dhanapala, Jayantha, "Nuclear Weapon Free Zones: Challenges and Opportunities", *Development and Dialogue*, 2001, 1-2, p.18.
6. See *Official Record of the Disarmament Commission*, Supplement for January-December 1956, UN Doc. DC/33, Annex 5.
7. For the details of the Repacki Plan see, "A World at Peace", *Report of the Palme Commission on Disarmament and Security*, Stockholm, April 1989.
8. Pande, Savita, "Tlateloco Treaty: How Successful?" *Strategic Analysis*, Vol.22, No.1, April 1998, p.38.
9. Press Release of the George Town Summit of the NAM.
10. Block, L., "Great Power Intervention and Conflict Potentials in the Third World", *Bulletin of Peace Proposals*, Vol.13, No.4, 1982, p.115.
11. UN General Assembly Resolution 3261 F(XXIX), 1974.
12. *The Question of Nuclear Weapon Free Zones in all its Aspects*, Special Report of the CCD (New York: United Nations, 1976).

2

Concept and Value of the Nuclear-free Zones

NUCLEAR WEAPON-FREE ZONES: THE CONCEPT

The 1975 Special Session of the United Nations General Assembly defined the nuclear weapon-free zones in the following words:

I. A nuclear weapon-free zone shall, as a general rule, be deemed to be any zone, recognised as such by the General Assembly of the United Nations, which any group of states in the free exercise of their sovereignty, has established by virtue of a treaty or convention whereby:

   a) The statute of total absence of nuclear weapons to which the zone shall be subject, including the procedure for the delimitation of the zone, is defined;

   b) An international system of verification and control is established to guarantee compliance with the obligation deriving from that statute.[1]

II. In the same resolution, the General Assembly set out the principal obligations of nuclear-weapon states towards such zones and the states they included. It stated that "In every case a nuclear weapon-free zone, recognised as such by the Assembly, all nuclear weapon states shall undertake to reaffirm, in a solemn international instrument having full legally binding force such as a treaty, a convention or a protocol. These obligations are:

a) To respect, in all its parts, the statute of total absence of nuclear weapons defined in the treaty or convention which serves as the constitutive instrument of the zone;

b) To refrain from contributing, in any way, to the performance in the territories forming part of the zone of acts which involve a violation of the aforesaid treaty or convention; and

c) To refrain from using or threatening to use nuclear weapons against the states included in the zone.[2]

In 1978 the issue was again referred to the General Assembly which further elaborated the concept to include Weapons of Mass Destruction also (it included nuclear, biological, chemical and radiological weapons). Apart from the definition provided by the United Nations, a number of scholars have also described the concept of nuclear weapon-free zone. For example, Ramesh Thakur, an expert on the South Pacific, is of the opinion that the concept of nuclear weapon-free zone arose from the desire of the non-nuclear countries to disengage on their own terms, from the nuclear madness of nuclear powers. It denotes a zone "which a group of states may establish by a treaty whereby the status of total absence of nuclear weapons to which the zone shall be subject, is defined and a system of verification and control is set up to guarantee compliance."[3]

Another scholar, Helen Leigh Phippard, holds that nuclear weapon-free zone has a particular role to play in those regions, free of nuclear weapons where they can prevent nuclearisation. Nuclear weapon-free zones can promote the security of non-nuclear weapon states both by obtaining pledges from nuclear weapon states regarding non-use of weapons against them and by discouraging or preventing the deployment of nuclear. weapons within their own region.[4]

Danfulani Shuaibu Ahmad, while writing about nuclear weapon-free zones opines that the expression nuclear weapon-free zone suggests the necessity of a treaty signed between states within the same region with the aim of preventing the introduction of any form of nuclear weapons within the zone of regional demarcation either by states themselves as party to the treaty or by other foreign powers.[5]

Gabriele Kittle and others, in an article, have written that "nuclear weapon-free zones are setup with two main aims. While the first aim is to restrict the option held by the nuclear weapon states enabling them to wage a nuclear war, the second is to prevent the emergence of new nuclear weapon states." Further, "The concept of nuclear weapon-free zones could represent a form of regional denuclearisation attainable within the intermediate future, the long-term goal being comprehensive global denuclearisation."[6]

It can be derived from the above definitions that, broadly speaking, nuclear weapon-free zone is a method through which a particular region can be kept free of the nuclear weapons. The objective is not only to prevent introduction of nuclear weapons in a particular region, but also to denuclearise that region, if nuclear weapons are already present there. The ultimate goal of establishing more and more such zones is complete eradication of nuclear weapons from the face of the earth. In this way, nuclear weapon-free zones are a step ahead of Nuclear Non-Proliferation Treaty (NPT), since they not only prevent horizontal proliferation but also geographical proliferation.[7]

After understanding the concept of nuclear free zones or nuclear-weapon-free-zones the question that arises is that what would be the requirements of such a zone or what would be the principles governing such a zone. The Special Report of the Conference of the Committee on Disarmament entitled *Comprehensive Study of Nuclear Weapon Free Zones in All Its Aspects*, published by the United Nations in 1976, contains a detailed description. The study tried to prepare a broad framework of the requirements of a nuclear weapon-free zone (further discussion on UN Resolution will be based on that). However, it was very difficult to arrive at a consensus on every issue and the experts made it very clear that "it was not possible and realistic to set up precise guidelines for the creation of such zones", as the conditions are bound to differ from region to region.[8]

None-the-less, the main issues which the experts tried to deal with were: procedure for the establishment of the zone; activities related to nuclear weapons; obligations of the state parties to the treaty and the obligations of nuclear weapon states; verification and control measures, etc. In order to understand the concept of nuclear weapon-free zones, it is also necessary to consider some of the basic principles laid down for such zones.

PROCEDURE FOR THE ESTABLISHMENT OF NUCLEAR WEAPON-FREE ZONES

So far as the procedure for the establishment of a nuclear weapon-free zone is concerned, the first criteria, according to experts, is that "the initiative to create nuclear weapon-free zones should come from within the region." It is expected that all the countries of that region should participate in it and the treaty declaring that region as nuclear free should be of an 'unlimited duration'[9].

ACTIVITIES RELATED TO NUCLEAR WEAPONS

At the outset, the experts felt the necessity for a clear definition of the fundamental concepts underlying the idea of a nuclear weapon-free zone, such as 'Nuclear Weapons', 'Territory', etc. Secondly, regarding 'development and production of nuclear weapons', it was agreed by most of the experts that "the states of the zone should undertake not to develop, test or produce, by any means, whatsoever, any nuclear explosive device, either directly or indirectly, on behalf of themselves or anyone else, or in any other way." Further, the experts also agreed that if the territories in the jurisdiction of extra-zonal states are "brought within the scope of the zone, their undertaking of the same effect should become operative in those territories." Thirdly, another obligation which the states were expected to assume was "not to acquire, possess, or receive nuclear weapons in any manner, directly or indirectly, by themselves or by anyone on their behalf, or in any other way, inside or outside the zone." Fourthly, state members were required to undertake to prohibit and prevent, within territories under their jurisdiction, the installation and deployment, storage and stockpiling of nuclear weapons that are under the control of states outside the zone. And finally, the question relating to 'Transport and Transit of Nuclear Weapons' through the zone, which was one of the most controversial issues, was also discussed by the experts. But they failed to reach on any decision because of its contradiction with the Law of the Sea.[10]

UNDERTAKINGS *VIS-A-VIS* THE ZONE

In connection with the undertaking to be entered into, with the establishment of nuclear weapon-free zones, the states may be divided into three groups: those within the zone; nuclear-weapon states; and other states.

a) Undertakings by States within the Zones

The states belonging to the zone are expected to "ensure specifically that they will not develop, produce, possess or acquire nuclear weapons in the territory under their jurisdiction and that they will not permit anybody to store, install or deploy such weapons in their territories. In addition, they should agree not to give, seek or receive any assistance in the development and production of nuclear weapons."

b) Undertakings by Nuclear Weapon States

Most experts felt that the nuclear weapon states should respect the nuclear weapon-free status of the zone and "ensure not to use or threaten to use nuclear weapons against any state included in a nuclear weapon-free zone", which means they should undertake the following:

i) Not to install, deploy or stockpile nuclear weapons in the zone and if they have done so in the past, to withdraw them from the zone immediately;

ii) To guarantee that if they have military bases in the zone, they contain no nuclear weapons or probably should allow the bases or establishments to be inspected;

iii) Not to provide the states of the zone any assistance in the development, production or acquisition of nuclear weapons; and

iv) To cooperate in the establishment of a safety area adjacent to the zone, if special circumstances exist.

c) By Other States

The undertakings by non-nuclear, extra-zonal states are "not to carry out any activity endangering the efficient functioning of the zone and, in particular, not to provide any assistance to the members of the zone which might lead to the development or production of nuclear weapons."[11]

PEACEFUL USE OF NUCLEAR ENERGY

Most of the experts believed that "the benefits of peaceful use of nuclear energy, including potential benefits from peaceful nuclear explosions, should be available to state members of nuclear weapon-free zones. But, they could not reach at any concrete decision, as some experts held the view that because of the technological similarity in nuclear weapons and nuclear explosive devices for peaceful purposes (as it could not be detected in India, South Africa, etc.) the possession of such devices by zonal states would significantly weaken a zonal regime."[12]

RELATIONS WITH EXISTING SECURITY TREATIES

It is understood that treaties establishing nuclear weapon-free zone should be consistent with the other treaty obligations of the zonal states. "Establishment of a nuclear weapon-free zone should not interfere with the existing security arrangements", but the situation becomes complex, if any nuclear weapon state is a party to any such existing alliance. Matters can be further complicated, if the treaty provides for deployment of nuclear weapons in the territory of the zone party. In this regard, the experts argued that such cases should be dealt separately for each party.[13]

VERIFICATION AND CONTROL

A whole chapter in the Special Report has been devoted to the question of verification and control. In brief, it says that the viability of the nuclear weapon-free zones will largely depend on an effective system of verification and control that ensures the nuclear weapon-free status of the zone. This will promote confidence among the states of the zone which should also include the application of appropriate International Atomic Energy Agency (IAEA) safeguards that call for periodical inspections. In this regard, the role of IAEA safeguard system was specifically emphasised. However, the precise scope and nature of such a system would vary from zone to zone.[14]

RELATIONS WITH INTERNATIONAL LAW

Most of the experts believed that the creation of the nuclear weapon-free zones should be effected in accordance with the international law and with the principles of the Charter of the United Nations. In particular with the following fundamental principles guiding the mutual relations of states: sovereign equality and respect for the rights inherent in sovereignty; refraining from the threat or use of force against the territorial integrity or political independence of any state; inviolability of frontiers; territorial integrity of states; peaceful settlement of disputes; non-intervention in internal affairs; equal rights and self-determination of peoples; right of self-defence; cooperation among states, and fulfilment in good faith of obligations under international law.

In this regard, the experts were also of the view that, as a basic principle, the boundaries and the provisions of nuclear weapon-free zone should also be determined in accordance with international law including the principles of freedom of navigation on the high seas and in straits used for international navigation and international air space. It was further emphasised that this basic principle applies to the right of innocent passage through the territorial sea. Some other experts, however, emphasised that the provision of a nuclear weapon-free zone treaty should prohibit the transit of nuclear weapons through the territory of the zone including the entry into ports situated in the zone of vessels having nuclear weapons on board. Still some other experts pointed out that the matter should be referred to individual nuclear weapon-free zone treaties.[15]

It may, however, be noted that there is a significant difference between applying arms control measures in sea areas as compared to land areas. The United Nations Convention on the Law of the Sea (UNCLOS) which was signed on 10 December 1982 and entered into force on 14 November 1994 has been serving almost like a 'Constitution' for the countries on the matters related to sovereignty over seas. UNCLOS entitles all the states to use the "freedom of the high seas". But the Convention also prescribes that "the high seas shall be reserved for peaceful purposes" (UNCLOS article 88) and that "States shall refrain from any threat or use of force against the territorial integrity or political independence of any states, or in other manner inconsistent with the principles of international law embodied in the Charter of the United Nations" (UNCLOS article 301). Coastal states enjoy full jurisdiction only over their internal water. In exclusive economic zones or on the high seas the coastal states have no political rights. Zonal commitments applying to sea areas should, therefore, be prescribed separately by each zone treaty. Denuclearisation of a sea area would require agreement, in principle, among all states or at least among the nuclear weapon states to make the regime effective.[16] The transit issue was extensively debated when the Latin American zone was being negotiated and transit was left to individual zonal states to permit or deny in each case.

TERRITORIAL DISPUTES

Sometimes a territorial dispute between prospective zonal states or between a prospective zonal state and an extra-zonal state can complicate accession to a zone agreement. It is obvious that a state assuming treaty commitments in relation to its entire territory could not implement such commitments in any part of its claimed territory which is, *de facto*, controlled by another state. It may also be difficult for a state to assume zonal responsibilities for areas under firm national control but claimed by others. Dispute between Cuba and the United States due to which Cuba has not signed the Treaty of Tlatelolco is a good example of territorial dispute.[17]

Thus, the study shows that CCD members dealt at length with every possible issue related with nuclear weapon-free zones, although they could not arrive at consensus on all the points leaving some of them open-ended. Most of the aspects were left to the states of particular region desiring to declare their area as nuclear-free, to be interpreted according to their requirements. This particular fact has made the concept of nuclear weapon-free zone very loose and weak. As a result, effective implementation of these zones has not been possible. However, in spite of all the weaknesses, it cannot be denied that, if implemented properly, nuclear weapon-free zones will prove to be an effective method of nuclear non-proliferation.

THE VALUE OF NUCLEAR WEAPON-FREE ZONES

The former Soviet President, Mikhail Gorbachev, speaking at one of the international meetings on nuclear weapon-free zones, emphasised Soviet support to the concept and said, "the value of nuclear weapon-free zone lies in the contribution which they can make towards limiting the sphere of nuclear preparations, easing apart the nuclear forces of the alliance which confront one another, thereby reducing the risk of a nuclear conflict."[18]

Thus, generally it can be said that the establishment of nuclear weapon-free zone is an attempt to restrict the option of nuclear war by nuclear weapon states. In other words, they can play an important role in 'crisis prevention' and crisis management. Further, in the event of the actual hostility breaking out, they can make it possible to limit the destruction.[19] Nuclear weapon-free zones can also act as buffer zones between two crisis-ridden areas. During the Cold War period, a proposal to declare Central Europe as a nuclear weapon-free zone had come forward with the aim of making it a buffer zone between East West conflict.[20]

A few other scholars like Donna J. Click argue that if a conflict breaks out between two nuclear powers near a nuclear weapon-free zone, "the incentive to launch a first-strike would be reduced, since neither country would have battlefield or theatre nuclear weapons deployed in that area."[21] Moreover, the 'passive' (negative) guarantee given by the nuclear powers to the states of the zone is another important factor. This would ensure that the nuclear states would not use nuclear weapons against the zone. Such guarantees would make the territory immune to nuclear strike.[22]

Preventing Horizontal Proliferation

The basic aim of establishing nuclear weapon-free zones is to limit the spread of nuclear weapons by preventing their introduction into areas as yet free from them. Nuclear weapon-free zones have come to be recognised as most practical means to halt the horizontal proliferation.[23] By signing the nuclear-weapon-free zone treaty the parties guarantee not to produce or possess nuclear weapons. This can help contain regional arms race and denuclearise regional conflict.[24]

The breakdown of bipolar world and creation of several centres of power has created a situation where there is a great threat of use of nuclear weapons in the event of a regional conflict. In this respect, the concept has become more relevant in the present situation. The growth of technological know how in Third World countries has further increased this risk. One cannot ignore the fact that many countries would wish to acquire nuclear weapons in order to attain regional supremacy, such as the case of India and Pakistan.

Preventing Geographical Proliferation

Establishment of nuclear weapon-free zone in several areas of the world will also place a limitation on the geographical proliferation of nuclear weapons.[25] Apart from this, since the nuclear weapon-free zone secures a guarantee from nuclear weapon states to withdraw their nuclear bases from the declared zone and also restrict the transit of nuclear-powered ships, it would be able to successfully restrict geographical proliferation. In this respect, nuclear weapon-free zones are one step ahead of NPT.[26]

Confidence-Building Measures

Another significance of nuclear weapon-free zone lies in the fact that it could become a major confidence-building measure for any region.[27] Once ensured that the area is free of nuclear weapons, the nuclear arms race will automatically come to an end. It would also diffuse regional tensions and instability that increase the possibility of nuclear options. Moreover, once the process is started, it can be applied to non-nuclear weapons also, reducing the overall arms race.

Global Elimination of Nuclear Weapons

The establishment of nuclear weapon-free zone can be considered as a process in the direction of overall goal of global elimination of nuclear weapons. In this way, the ultimate goal can be achieved in stages.[28] Thus, it can be described as an important first step towards complete elimination of nuclear weapons. Therefore, these zones should be seen as a stepping stone rather than an end in themselves.[29]

In conclusion, with respect to the importance of nuclear weapon-free zones it may be stated that it has a limited but a significant role to play. Its desirability cannot be overemphasised, since it is just a beginning and not an end in itself. Creation of such a zone would always be considered a partial step and a contribution to the process eventually leading to general nuclear disarmament. To quote United Nations General Secretary, who has reflected similar ideas stating that "in a period where the widespread knowledge of nuclear technology and huge quantities of fissionable material are becoming available, nuclear weapon-free zones provide the best and the easiest means whereby non-nuclear weapon states can, by their own initiative and effort, ensure the total absence of nuclear weapons from their territories and enhance their mutual security. They can also provide a logical basis for promoting the peaceful uses of nuclear energy within the zone. Above all, they can provide the means for obtaining security assurances from the nuclear powers never to use or threaten to use nuclear weapons against Member States of the zone."[30]

Other than the above-mentioned features, since the initiative to establish such a zone comes from within the region, it becomes an significant gesture, as it is a kind of assertion of independence on the part of the regional countries (sometimes may be as small as Kiribati and Vanuatu). Reflecting similar sentiments in context of the Treaty of Tlatelolco, the Secretary General of the Agency for the Prohibition of Nuclear Weapons in Latin America (APANAL), Enrique Raman Morey has written, "From my point of view, the sovereign decision of the Latin American countries to declare themselves as the first nuclear weapon-free zone in a densely inhabited region of the planet, had no other significance than affirming their independent foreign policy and served their most explicit expression of opposition to foreign intervention in the region. This is an element that should be constantly praised by all researchers of international relations."[31]

This is an important aspect, if regions like the South Pacific, Southeast Asia, Africa and Latin America are taken into consideration where most of the countries have a colonial past and come into the category of either developing or least developed countries.

Moreover, even if the region where such a zone is created is already free of nuclear weapons or where all the countries have already adhered to Non-Proliferation Treaty, which has been highlighted by several critiques, than also it has a value. In the words of Greg Fry, a noted scholar on the South Pacific, "These treaties do not change any existing nuclear weapon activity, because there is none in these environments. The idea is to prohibit weapons involvement before it happens as in the case of Antarctica, Seabed, etc."[32]

This point again emphasises that even if such zones do not bring any revolutionary change in the overall spread of nuclear arms they cannot be termed as a 'futile exercise' as has been done by some scholars. In this regard, it will be relevant to discuss the existing nuclear weapon-free zones other than the Southeast Asia and the South Pacific which are the focus of this study.

REFERENCES

1. *UN Resolution 3472 B* (XXX), 1975.
2. Ibid.
3. Thakur, Ramesh, "Nuclear Weapon-Free Zone for the South Pacific: A New Zealand Perspective", *Pacific Affairs*, Vol.58, No.2, Summer 1985, p.217.
4. Phippard, Helen, Leigh, "Nuclear-Weapon Free Zones: Problems and Prospects", *Arms Control*, Vol.19, No.2, August 1993, p.93.
5. Ahmad, Daufulani, Shuaibu, "The Changing Geostrategic Theatre and Zone of Peace and Nuclear Weapon Free Zones: A Review", *Scandinavian Journal of Development Alternatives*, Vol.II, No.2, June 1992, p.65.
6. Kittel, Gabriele and others, "Nuclear-Weapon-Free Zones: Reevaluating a Political Concept of Peace Making", *Bulletin of Peace Proposals*, Vol.22, No.2, June 1991, p.222.
7. Phippard, n.4, p.93.
8. Special Report of the CCD, p.54.
9. Ibid., pp.34-35.
10. Ibid., pp.36-39.
11. Ibid., pp.40-41.
12. Ibid., pp.52-54.
13. Ibid., p.50.
14. Ibid., pp.43-47.
15. Ibid., pp.48-50.

16. *A Zone Free of Weapons of Mass Destruction in the Middle East*, United Nations, 1996, p.50.
17. Ibid., p.45.
18. A Documentary Account of International Meeting for Nuclear-Weapon-Free-Zone, Berlin, 20 June 1988, *Pakistan Horizon*, Vol.43, No.1, 1990, p.71.
19. Click, Donna J., "A Balkan Nuclear Weapon-free Zone", *Journal of Peace Research*, Vol.24, No.2, June 1987, pp.111-12.
20. Kittel, et. al., No.6, p.217.
21. Click, n.19, p.111.
22. Multan, Wojciech, "The Past and Future of Nuclear Weapon Free Zones", *Bulletin of Peace Proposals*, Vol.16, No.4, 1985, p.379.
23. Duah, Balfour, Agyemen, "Nuclear Free Zones and Disarmament", *Africa Today*, Vol.32, Nos.1-2, 1985, p.79.
24. Kittel et. al., no.6, p.220.
25. Phippard, no.4, p.93.
26. Kittel, et. al., no.6, p.218.
27. Based on personal interview with Dr. Ian Talbot, School of International Studies and Law, Coventry University, London.
28. Wolfsthal, John, Brook, "Nuclear Weapon Free Zone: Coming of Age", *arms Control Today*, Vol.23, No.2, March 1993, p.4.
29. Jayantha, Dhanapala, "Nuclear Weapon Free Zone: Challenges & Opportunities" *Development and Dialogue*, 2000, 1-2, p.5.
30. *Introduction to the Report of the Secretary General on the Work of the Organisation*, 30th Session, Supplement No.1A (A/10001/Add.1), Sect.VIII.
31. Raman-Morey, Enrique, "Precursor of Other Nuclear-Weapon-Free Zones" in Pericles Gasparini Alves and Daiana Belinda Cipollone (ed.), *Nuclear-Weapon Free Zones in the 21st Century*, United Nations Institute for Disarmament Research, 1998, p.9.
32. Fry, Greg, "The South Pacific Nuclear-Free Zone: Significance and Implications", *Bulletin of Concerned Asian Scholars*, Vol.18, No.2, April-June 1986, p.68.

3

The Existing Nuclear-free Zones

NUCLEAR WEAPON-FREE ZONES IN UNINHABITED AREAS

Multilateral agreements which are supposed to contribute to establishing nuclear-free zones include Antartica Treaty (1959), the Treaties on outer space and on the moon (1967/1979) and the Seabed Treaty. Their importance lies in their endeavours to exclude specified areas from becoming nuclear armed. They have prevented the deployment of nuclear weapons and other weapons of mass destruction in the specific areas.

The Antarctica Treaty was the first international agreement establishing a demilitarised zone. The treaty establishes that Antarctica should be used for peaceful purposes only. (Article I) It also prohibits any nuclear explosion and disposal of radioactive waste material (Article 5). The treaty not only intended to solve the problem of different territorial claims but to ensure access to the whole of the area in order to carry out scientific research (Article III and IV).[1]

The United Nations has considered the question of outer space since the very beginning of the space age in the late 1950s. As early as 1957, proposals were made in the Disarmament Commission to establish an inspection system that would ensure that the launching of the objects through outer space would be exclusively for peaceful and scientific purposes. The treaty provisions are undertaken by the parties not to place in orbit around the earth any other kind of weapons of mass destruction, install such weapons on celestial bodies or station such weapons in outer space (Article 4). It also prohibits all military activity on the moon and other celestial bodies (article 4) and that all stations, installations, equipment and space vehicles would be open to state parties (Article 12).[2]

The Seabed Treaty also contains a prohibitive regulation which is directed against nuclear technical facilities and weapons of mass destruction. Under article I parties of the treaty undertake not to implant, or emplace on the seabed and the ocean floor and in the subsoil thereof, any nuclear weapon or any other types of weapons of mass destruction. Article II states that outer limit of the seabed zone should be 12 miles territorial sea limit. Article III provides for an verification procedure which involves observation of activities in the sea bed zone to be followed in the event of suspected violation.[3]

Apart from these three treaties concerning extra-territorial areas, several nuclear free zones have come up in inhabited areas as well and these are the Treaty of Tlatelolco, Treaty of Pelindaba, the Rarotonga Treaty and the Bangkok Treaty.

THE TREATY OF TLATELOLCO

The Treaty for the prohibition of nuclear weapons in Latin America (Treaty of Tlatelolco) has been hailed as the first treaty to establish nuclear weapon-free zone in an inhabited area. It is also the first agreement to establish a system of International control, and a permanent supervisory organ, the Agency for the Prohibition of Nuclear Weapons in Latin America (APANAL). It was opened for signatures in February 1967 and entered into force in April 1967. The treaty was born in the aftermath of the Cuban missile crisis when the world was brought to the brink of a nuclear war. The governments and the people of Latin America and Carribean decided that never again should they be subjected to the threats and fears of a nuclear wars in their respective territories.[4] It originated from the Brazilian idea and covers around 20 million sq. km of territory inhabited by around 350 million Latin American people. The treaty has been signed by 32 states as of 2001.[5]

The basic obligations of the treaty are contained in Article I in which the contracting parties undertake not to test, use, manufacture, produce or acquire nuclear weapons for themselves or for anyone else; not to aid or encourage others in such conduct and not to receive, store, deploy or otherwise possess nuclear weapons. Article 7 of the treaty establishes OPANAL which consists of a General Conference of all parties with equal voting power. Annexed to the treaty are two additional protocols. Under Protocol I, France, the Netherlands, the United Kingdom and the United States would agree to guarantee nuclear weapon-free status to those territories, for which they are 'de jure', 'de facto', internationally responsible. All the parties have signed this protocol except France. Under Protocol II, negative security guarantees by the nuclear states has been taken. They pledge to respect fully the "denuclearisation of Latin America in respect of war-like purposes" and "not to use or threaten to use nuclear weapons against the contracting parties." Protocol II has been accepted by all the nuclear weapon states but with some reservations.[6]

The Treaty of Tlatelolco, although a commendable step in the sense of being a role model for regional efforts of denuclearisation, has in reality achieved little in this regard. The troublesome issue with regard to effective implementation of the treaty was the stand taken by the two most important countries of the region, i.e., Argentina and Brazil. Both the countries had refused to sign NPT and were committed to retain the freedom to develop nuclear technology in future. They differed from other parties to the treaty on most of the issues related to the treaty. They held the view that peaceful nuclear explosions were permitted to the parties. They also differed from the rest of the regional countries on transportation, reservations, etc. Finally, it took 26 years for Argentina and 27 years for Brazil to formally become parties to the treaty. It became possible only after the signing of bilateral agreement between the two countries at Guadalaraja in 1990. This, coupled with change of governments in these countries and economic constraints led them to give up their nuclear weapons plans.[7] Thus, the full implementation of the treaty in the foreseeable future is, therefore, a genuine possibility.

AFRICAN NUCLEAR WEAPON-FREE ZONE TREATY (TREATY OF PELINDABA)

The proposal for establishment of nuclear weapon-free zone in Africa dates back to the early sixties after the beginning of French nuclear tests at Sahara.[8] The proposal was also taken to the United Nations General Assembly. However, in 1964 a declaration on denuclearisation of Africa was issued by the Organisation of African Unity (OAU). Later on, 28 African states submitted the proposal in the United Nations which was endorsed by the General Assembly. However, it was again during the 30th Session held in 1975 that a new resolution was adopted with some new elements. It may be noted in this regard that by this time France had already shifted its test site to the Pacific but African states were concerned about the growing capabilities of South Africa in the field of nuclear energy.[9] The South African nuclear policy remained the greatest hurdle in the conclusion of the treaty on denuclearisation of Africa. It was only after the dismantling of South African nuclear programme and its accession to NPT in 1990 that the Treaty of nuclear weapon-free zone in Africa could take shape. The draft for the treaty was finalised by the 31st Session of the OAU Assembly, and the treaty was opened for signatures on 11 April 1996.

The treaty prohibits the research, development, manufacture and acquisition of nuclear explosive devices. Each party remains free to allow visits, as well as transit by foreign ships and aircraft (Article 4). The treaty also prohibits any attack against nuclear installations. The parties undertake not to dump or permit the dumping of radioactive wastes and other radioactive matter anywhere within the zone (Article 7). The parties should conclude an agreement with the IAEA for the application of comprehensive safeguards to their peaceful nuclear activity (Articles 8 and 9). Under Article 2 the treaty establishes the AFCONE, consisting of 12 members elected by the parties, for the purpose of ensuring compliance with its provisions. The zone includes the territory of the continent of Africa, island states members of OAU and all the islands considered by the OAU to be part of Africa. Annex I of the treaty contains the map indicating the zone of application.[10]

Other than these provisions there are three protocols of the treaty. Under Protocol I, all the five recognised nuclear weapon states are to undertake not to use or threaten to use a nuclear explosive device against the parties to the treaty (this is signed by all the five countries but ratified only by China and France). Under Protocol II, the five nuclear powers are to undertake not to test nuclear explosive devices anywhere within the zone (this protocol is signed by all the five countries but ratified only by China and France). Under Protocol III, states with territories within the zone for which they are internationally responsible are to undertake to observe certain provisions of the treaty with respect to these territories. It was addressed to France and Spain. All the three protocols, although have been signed by all the nuclear weapon states yet they have declared that they were not bound by it, in case of any invasion (Protocol I).

Article 18 of the treaty envisages that the treaty will come into force after the 28th ratification. With Kenya depositing its instrument of ratification in January 2001, up to now 15 ratifications have been deposited.[11] However, as far as the value of the treaty is concerned it may be mentioned that it came into existence only after South Africa became a party to NPT and the demise of the cold war and related fears. Therefore, it is to be seen in future what the treaty can achieve in terms of denuclearisation of the region.

THE PROPOSAL FOR NUCLEAR WEAPON-FREE ZONE IN SOUTH ASIA AND INDIAN STAND THEREON

The idea for establishing a nuclear weapon-free zone in South Asia was mooted by Pakistan and was taken up by the United Nations General Assembly in 1974. India and Pakistan submitted separate draft resolutions in the first committee. The Indian draft held the view that the creation of a nuclear weapon-free zone in the appropriate region of Asia should come from the states of the region concerned. Whereas, the Pakistani resolution suggested that the concept should be endorsed first and then, the concerned states should be invited to initiate the process. India also held that South Asia could not be considered a distinct zone, as it was an integral part of the Asian and Pacific region, which was surrounded by nuclear weapon states, or countries belonging to their alliances and also that India had no intention to develop nuclear weapon and would use the nuclear energy (after Pokharan explosion) for peaceful purposes.[12] The Indian draft which eventually became Resolution 3265A obtained one hundred and four votes with twenty seven abstentions and Pakistani draft which became Resolution 3265B obtained ninety six votes with thirty six abstentions. India voted against it.[13] As far as nuclear weapon states are concerned, the United States, in the United Kingdom and France abstained from voting, USSR voted in India's favour and China voted in Pakistan's favour.[14] In the 1975 session, the United Nations General Assembly once again adopted Resolutions both by India and Pakistan which were a repetition of the previous year.[15] The issue was taken again and again in the United Nations. However, in 1977 the Indian resolution was dropped with its consent but Pakistan continued with its initiatives in the United Nations.

Outside the United Nations, Pakistan also took the issue to several fora multilaterally, such as Islamic Summit, ASEAN etc. and with Canada, Sri Lanka bilaterally. Pakistan in this regard stressed that there was no difference between peaceful nuclear explosion and one that would develop a nuclear weapon. In reality, what Pakistan was trying to do was to create a positive 'image' for itself as a 'peace seeker' and to win a diplomatic victory over India which initially it did manage to achieve but very soon, the proposal became a dead letter.

India on its part, always remained firm on its stand of rejecting the proposal on the following ground; first, creating a nuclear weapon-free zone is a matter that has to be first resolved by the regional powers themselves and that the proposal should come from within the region (a criterian accepted by the UN General Assembly itself); Second that India had no intentions to develop nuclear weapons; third that the idea of nuclear weapon-free zone is unrealistic and that the movement and deployment of nuclear weapons in various regions of the world by nuclear weapon states are fundamentally irreconcilable with the very idea of nuclear weapon-free zone. At the second special session of the General Assembly on disarmament held in 1982, India's Foreign Minister stated that India could not subscribe to the legitimisation of the possession of the nuclear weapon by a few powers by agreeing to live under their professedly benign protection in the guise of a nuclear weapon-free zone. And fourth, India also maintained that nuclear disarmament was a global and not a regional issue. Security provided by nuclear weapon-free zone was illusory, given the global reach of nuclear weapons.[16] In reality, for India its security equations with China have been the issue of major concern with regard to nuclear weapon-free zones in South Asia. China has been perceived by India as its primary strategic threat and proposal for nuclear weapon-free zone without China being a party to it could have been never acceptable for India.

With the Pokhran tests of May 11, 1998 by India followed by Pakistan, the proposal for South Asian nuclear free zone which was already an almost dead proposal was blown to pieces. Further, Pakistan's refusal to reciprocate India's 'no first use' pledge makes the situation all the more alarming. In this scenario, any treaty or convention which could stop the nuclear proliferation by the two countries and at least bring the matters to a *status quo* would be a welcome step.

After discussing the individual cases of nuclear weapon-free zones it is important that the subject be viewed with respect to issues like attitude of nuclear weapon states and NPT, for a better understanding.

REFERENCES

1. See Text of the Treaty in *SIPRI yearbook of Armament, Disarmament and International Peace*, 1973, pp.487-493.
2. See Comprehensive Study of Nuclear Weapon Free Zone, pp.11-12.
3. See Text of the Treaty in SIPRI Yearbook of Armament, Disarmament, 1972, pp.537-541.
4. William Epstein, "Tlatelolco and the Nuclear Weapon Free World" in Pericles Gasparini and others, (eds)., *Nuclear Weapon Free Zones in the 21st Century* (Geneva: United Nations Institute for Disarmament Research, 1997), p.23.
5. *SIPRI Yearbook of Armament & Disarmament*, 2001, p.652.
6. See Text of the Treaty published in Ibid.
7. For details on these issues, see Savita Pande, "Tlatelolco Treaty: How Successful?" *Strategic Analysis*, Vol.22, No.1, April 1998, pp.35-48.
8. Sola Ogunbanwo, "History of the Efforts to Establish an African Nuclear-Weapon Free Zone", *Disarmament*, Vol.19, No.1, 1996, p.14.
9. For the details on above discussion see Savita Pande, "Treaty of Pelindaba: How Different?" *Strategic Analysis*, Vol.22, No.4, pp.547-559.
10. See the Text of the Treaty published in *SIPRI*, 1997.
11. The details about the treaty have been taken from *SIPRI Yearbook*, 2001, p.666.
12. *The United Nations & Disarmament*, 1970-1975 (New York: United Nations, 1976), p.104.
13. See UN Doc.A/PV.2309.

14. Documents on Disarmament 1974 (Washington: United States Arms Control and Disarmament Agency).
15. Resolution 3476 (XXX) and 3476B(XXX).
16. See, Ross Masood Husain, "A Nuclear-Weapon-Free Zone in South Asia", *UN Regional Disarmament Workshop for Asia-Pacific* (New York: UN, 1991), pp.210-214.

4

Nuclear-free Zone *vis-a-vis* Nuclear Weapon States and NPT

NUCLEAR WEAPON-FREE ZONES AND THE NUCLEAR WEAPON STATES

In the United Nations study on nuclear weapon-free zones, clear obligations have been laid out for the nuclear weapon states with regard to nuclear weapon-free zone. The term 'Nuclear Weapon State' has been defined in Article IX:3 of the Non Proliferation Treaty, as a state having manufactured and exploded a nuclear weapon or other nuclear explosive devices prior to 1 January 1967. The definition does not cover a new country acquiring after the stated date and only recognises five nuclear powers, namely, United States, Britain, France, Russia and China.[1]

It is important for effective implementation of any nuclear arms control mechanism to gain support of the nuclear weapon states. In the case of nuclear weapon-free zones also the support of five nuclear power is critical for their effective operation. A passive or negative guarantee obtained from five nuclear powers is an important component of any nuclear weapon-free zone. The issue has also been taken up in the United Nations in several of its meetings. In 1978 all the nuclear weapon states made individual declarations designed to assure non-nuclear weapon states against the use or threat to use cf nuclear weapons. However, some differences existed among the assurances given by the five countries. The former Soviet Union, declared not to use nuclear weapons against states which have renounced the production and acquisition of such weapons and which do not have such weapons in their territory. In another statement Soviet Union renounced the first use of nuclear weapons. The stand taken by China in this matter was similar to that of the Soviet Union. It assured not to use nuclear weapons against non-nuclear states and also gave the guarantee for no first use. The Western countries although, gave a guarantee not to use nuclear weapons against such zones but laid down certain reservations. They exempted from their guarantee, the states allied with nuclear power that has committed an act of aggression against the territories of these powers and their allies. The three Western nuclear weapon states also did not commit for no first use.[2]

Nevertheless, the nuclear weapon states have acknowledged in principle the significance of nuclear weapon-free zones as a means of preventing nuclear proliferation. Their main interest, of course is to see their own concept of horizontal non-proliferation implemented. Almost all the nuclear weapon countries have laid down specific criteria of their own regarding the essentials of nuclear weapon-free zones and proposed them in the United Nations Disarmament committee meeting in 1997.

The USA have since long adopted seven general criteria that must be met as a condition for accepted and assisting a nuclear weapon-free zone (NWFZ).[3]

i) The initiatives for the creation of the NWFZ should come from the states in the region concerned;

ii) All states whose participation is deemed important should participate in the zone.

iii) The zone arrangement should provide adequate verification of compliance with its provisions;

iv) The establishment of the zone should not disturb existing security arrangements to the detriment of regional and international security or otherwise abridge the inherent right of individual or collective self-defence guaranteed in the United Nations Charter;

v) The zone arrangement should effectively prohibit its parties from developing or otherwise possessing any nuclear device, for whatever purpose.

vi) The establishment of the zone should not affect the existing rights of its parties under international law to grant or deny other states transit privileges within their respective land territory, internal waters, and air space to nuclear-powered and nuclear-capable ships and aircraft of non-port nations, including port calls and overflight; and

vii) The zone arrangement should not seek to impose restrictions on the exercise of rights that are recognised under international law, particularly the high seas, freedoms of navigation and over flight, the right of innocent passage of territorial and archipelagic seas, the right of transit passage of international straits, and the right of archipelagic sea lanes passage of archipelagic waters.

The European Union including as member states the two nuclear weapon powers, France and the United Kingdom, stated a similar view:[4]

i) A nuclear weapon-free zone should be based on arrangements freely arrived at among the states of the region concerned;

ii) The initiative for the creation of nuclear weapon-free zone should come from states within the region concerned;

iii) The nuclear weapon-free zone should form a coherent and clearly defined geographical entity;

iv) All states in the region should participate in the nuclear weapon-free zone;

v) A nuclear weapon-free zone should take into account the relevant specific characteristics of the region concerned.

vi) A nuclear weapon-free zone should not have adverse effects on existing regional or international security arrangements, nor on the inherent right to individual or collective self-defence as enshrined in the Charter of the United Nations, nor on regional or international stability;

vii) A nuclear weapon-free zone should provide for the effective prohibition of the development or possession of any nuclear explosive device for any purpose of state parties, and should provide for state parties not to allow the stationing in their territory of any nuclear explosive device;

viii) A nuclear weapon-free zone should be consistent with existing international instruments of nuclear non-proliferation and disarmament;

ix) A nuclear weapon-free zone should be drawn up in manner consistent, with international law and the undertakings sought from the nuclear weapon states;

x) The obligations of all states parties to a nuclear weapon-free zone should be clearly defined and legally binding;

xi) A nuclear weapon-free zone should provide for effective verification of the obligations assumed by its parties, i.e., by the application of full-scope international atomic energy safeguards to all peaceful nuclear facilities in the zone;

xii) A nuclear weapon-free zone should be in full accordance with international law, in particular with all the rights of transit and of free navigation and overflight on the high seas.

Finally, the position of China was also stated at the UNDC meeting:[5]

i) Nuclear weapon-free zones should be established by relevant countries in the light of realities of their region on the basis of voluntary agreement through consultations among themselves.

ii) Treaties on nuclear weapon-free zones should be consistent with the purposes and principles of the Charter of the United Nations and should not be used to interfere in the internal affairs of the countries outside of the relevant nuclear weapon-free zone.

iii) The nuclear weapon-free status of nuclear weapon-free zones should not be subject to influence of any other security mechanism. Countries in nuclear weapon-free zones should not refuse to fulfil their obligations under any excuses, including that of a military alliance.

iv) A nuclear weapon-free zone should have a clear geographical border. It should not include continental shelves and EEZs, nor the areas where there exist disputes over sovereignty of territory or maritime rights and interests between the contracting parties to the nuclear weapon-free zone treaty and their neighbouring countries.

v) Effective verification mechanism, including IAEA's safeguards, should be put in place in nuclear weapon-free zones so as to effectively prevent the proliferation of nuclear weapons.

vi) The arrangements of nuclear weapon-free zones should be conducive to the international cooperation in peaceful use of nuclear energy among member states so as to promote the development of their economy and science and technology.

vii) The nuclear weapon states should respect the status of nuclear weapon-free zones, undertake corresponding obligations and commit themselves to unconditional non-use and threat of use of nuclear weapons against nuclear weapon-free zones.

NUCLEAR WEAPON-FREE ZONES AND THE NPT

Several scholars view the concept of nuclear weapon-free zone as being complementary to Non-Proliferation Treaty. In view of this a comparison would be relevant here. The Non-Proliferation Treaty came into existence in 1968. Article VII of the treaty states, "Nothing in this Treaty affects the right of any group of states to conclude regional treaties in order to assure the total absence of nuclear weapons in their respective territories.[6] This implies that NPT in a way legitimises the concept of nuclear weapon-free zones. As far as horizontal proliferation is concerned, nuclear weapon-free zones certainly go by the spirit of NPT as they also ban the party to "manufacture, acquire, possess or have control over any nuclear explosive devices."

Article II of NPT also reflects similar sentiments as obligations on the part of non-nuclear states. As per the article they are not to receive the transfer of nuclear weapons or other nuclear explosive devices or control over them directly or indirectly, and not to manufacture or otherwise acquire nuclear weapons or other nuclear explosive devices and not to seek or receive any assistance in the manufacture of nuclear weapons or other nuclear weapon devices.[7] In this respect nuclear weapon-free zones reinforce the NPT.[8] This is the reason why almost all the review conferences of NPT have emphasised on the formation of nuclear weapon-free zones.

But nuclear weapon-free zone goes a step further than just preventing horizontal proliferation as it also prohibits deployment of nuclear weapon or nuclear explosive devices. In the words of Ramesh Thakur, "The concept is thus sharply different from the Nuclear Non-Proliferation Treaty of 1968 in one crucial respect, that is, states that do not possess nuclear weapons and adhere to the Non-Proliferation Treaty, still accept a stationing of nuclear weapons on their territories .... In contrast, nuclear weapon-free zone status prohibits such stationing."[9] Wojciech Multan and Helen Leigh-Phippard have also stressed this point. In the words of Multan, "The obligations arising under nuclear weapon-free zones are wider than the obligations arising under NPT,"[10] whereas, Phippard explains it as "nuclear weapon-free zones have a much broader disarmament objective than that of NPT. The NPT recognises the existence of five nuclear weapon states and tacitly permits them to station nuclear weapons on the territory of non-nuclear weapon states."[11]

Another major point of difference between NPT and nuclear weapon-free zones is of 'negative guarantee'. Although under NPT, nuclear weapon states have given a positive assurance but the non-nuclear weapon states have always expressed preference for a negative assurance that is a commitment by nuclear weapon states that they would not use nuclear weapons against them. The issue was the most contentious one and was taken up in all the review conferences of NPT,[12] whereas, in the case of nuclear weapon-free zones such an assurance is taken from the nuclear weapon states. Thus nuclear weapon-free zone as compared to NPT, represents a more far reaching political means of arms control since it embraces not only horizontal non-proliferation but also geographical non-proliferation. It also limits the options of nuclear weapon states by requiring them to provide negative security assurance.[13] Hence, the differences between the two are quite evident and nuclear weapon-free zone goes a step further than NPT.

ZONE OF PEACE

It is important to understand the concept of zone of peace as it is similar to the concept of nuclear weapon-free zone and also because in the case of Southeast Asia the nuclear weapon-free zone will be taken up as a component of a zone of peace, freedom and neutrality.

A zone of peace like a nuclear weapon-free zones is also regional initiative for disarmament. However, while nuclear weapon-free zones talk about denuclearisation, a zone of peace is a much more broader concept which talks about demilitarisation. Nuclear weapon-free zone stands for security against nuclear weapons, a zone of peace talks about overall security system of a particular region.

According to Danfulani Shuaibu Ahmad, a 'Zone of Peace' is an expression founded on the establishment of a 'strategic climate' of peaceful co-existence among nations, partly to an agreed treaty aimed at dissuading the component states from engaging in any form of activity capable of disrupting inter-state relations, regional peace and common security. He further explains that the concept suggests that no state within the region, be it an economic or a military power or both, should constitute a threat to the survival of others.[14] George W. Shepherd has described a Zone of Peace as a strategy of self-reliance which leads to delinkage from security and corporate systems of major powers. The idea is that a group of nations can exist more peacefully if the intrusion of the external powers is limited, especially in the security realm.[15]

From the above definitions, it can be gathered that the Zone of Peace is an arrangement which can be reached by an international agreement for a particular region in which it is expected that the arms race will be reversed, external threats will be reduced, and a new opportunity of peaceful settlement of disputes will be created. It also stipulates demilitarisation of that particular region. The very fact that a zone of peace is proposed for a particular region indicates that the region is exposed to great power rivalry, which has led to excessive militarisation of that area. The interference of external powers also has a tendency to manipulate the local conflicts in terms of their own interests.

Thus, it is necessary to delink from the great power rivalry at the first stage and then search for ways of collective self-reliance and peaceful settlement of disputes, and form a climate which promotes trade and development rather than arms race. While this process is essentially international, it can also be maintained at the regional level with the support of transnational groups. The areas proposed to be converted into zones of peace and the condition prevailing there shows that the external interference is the basic cause of their instability and excessive arms race. This is also true for the Indian Ocean and Southeast Asia for which the proposals have come forward.

The zone of peace is a relatively new concept. It first came up in the context of the Indian Ocean at the Lusaka Summit of Non-Aligned Nations in 1970 wherein one of the demands to be made to the Superpowers was to draw their attention towards ensuring regional peace and international security. The subsequent Non-Aligned Summits have also always raised this issue.[16]

The proposal was then adopted by the United Nations General Assembly in its Resolution 2832 (XXVI) of 16 December 1971. A special Indian Ocean Committee composed initially of 15 members was created to study the matter deeply.[17] The UN experts have not as yet succeeded in producing a definite framework of the concept. However, on the basis of the 1971 Resolution pertaining to the Indian Ocean as a zone of peace, a rough idea can be formulated with regard to the basic components of a zone of peace. These components are codes regulating the behaviours of regional states as well as external powers. These codes call upon the great powers to hold consultations with regional powers for "halting the arms race and expansion of their military presence" and "elimination from the area of all their military bases, installations, nuclear weapons and weapons of mass destruction." They also say that "warships and military aircrafts would not use the Indian Ocean for any threat or use of force."[18] But in the context of a Zone of Peace, Freedom and Neutrality in Southeast Asia, the concept has much relevance and that is the reason why the countries of the region are desirous of declaring Southeast Asia as a zone of peace, freedom and neutrality.

Finally, turning the issue to Southeast Asia and South Pacific, they present two different cases of nuclear weapon-free zones. Both the regions were exposed to the nuclear danger but in different ways. In the case of Southeast Asia, it remained an arena of Superpower rivalry throughout the cold war. The United States erstwhile Soviet Union and China (all the three being nuclear weapon states) tried endlessly to establish their supremacy bringing a hidden nuclear threat to the region. Whereas, in the case of the South Pacific, its relative isolation and remoteness made it an attractive site for nuclear testing and dumping of nuclear wastes. Consequently, the need for nuclear weapon-free zone was felt by the countries of both these regions.

REFERENCES

1. See test of the treaty published in SIPRI Year Book of Armament and Disarmament, 1969).
2. See *United Nations and Disarmament*, 1945-1985 (new York: United Nations, 1985) p.83.
3. UN Document A/CN. 10/1997/WG.I/WP. 11
4. http://www.uspid.dsi.unimi.it/proceed/cast97/prawitz.html
5. Speech by Head of the Chinese Delegation to the International Conference, 'Central Asia Nuclear Weapons Free Zone', Tashkent, Uzbekistan, 15 September 1997
6. *The United Nations and Disarmament*, no.2 p.78.
7. Ibid., p.78.
8. Hermann, Von Michael, "Europe Needs a Churchill of Peace", *Problems of Peace and Socialism*, November 1988, p.43.
9. Thakur, Ramesh, In *Defence of New Zealand: Foreign Policy Choices in the Nuclear Age* (London: Westview Press, 1986), p.149.
10. Wojciech, Multan,"The Past, Present and Future of Nuclear Free zones", *Bulletin of Peace Proposals*, Vol 16, No. 4, 1985, p. 222.
11. Helen Leigh Phippard, "Nuclear Weapon Free Zone: Problems and Prospects",*Arms Control*, Vol. 19, No. 2, August 1993, p. 93.
12. See United Nations and Disarmament, n.2.
13. Gabrial, Kittal and Others, "Nuclear-Weapon-Free Zones: Reevaluating a Political concept of Peace Making", *Bulletin of Peace Proposals*, Vol. 22, No.2, June 1991, p. 221.
14. Danfulani, Shuaibu Ahmed, "The Changing Geostrategic Theatre and Zone of Peace and Nuclear Weapon Free Zone: A review", *Scandinavian Journal of Development and Alternatives*, Vol. 11, No.2, June 1992, p. 65.
15. Shepherd, George W., "Developing Collective Self-Reliance in the Zone of Peace", *IDSA Journal*, Vol.15, No.1, July-September 1982, pp.19-20.
16. See Press Release of Lusaka Summit published in *Keesing's Contemporary Archives*, 1971.
17. Ahmad, No.14, p.67.
18. *The United Nations and Disarmament:* No.2, pp.104-5.

PART II

Nuclear Weapon–free Zone in Southeast Asia

5

Regional and Extra Regional Threats

On 15 December 1995, at a summit meeting of the Association of Southeast Asian Nations (ASEAN) held in Bangkok, the Southeast Asian Nuclear Weapon Free Zone Treaty (SEANWFZT), also known as the Treaty of Bangkok, was signed by all the countries of Southeast Asia, namely, Indonesia, Malaysia, Thailand, the Philippines, Singapore, Brunei, Vietnam, Laos, Cambodia and Myanmar. The Bangkok Treaty is a major initiative of ASEAN in its search for a new order in the region. The two most unique features of the treaty which set it apart from the rest of the nuclear-free zones are, first, the treaty also includes the continental shelves and 200 mile Exclusive Economic Zone (EEZ) of each state party and, second, in its protocol, the five nuclear weapon states are to undertake to refrain from using or threatening to use nuclear arms not only against parties to the treaty but also anywhere within the zone.[1]

The idea to establish a nuclear weapon-free zone in Southeast Asia was mooted in 1970s as a part of the proposal of Zone of Peace Freedom and Neutrality (ZOPFAN). The idea of ZOPFAN was a broader concept aiming to restrict external intervention as well as promoting regional cooperation. The nuclear weapon-free zone proposal aimed at limiting the great power intervention. The internal situation of Southeast Asia throughout the Cold War era was such that it had almost become a 'cockpit' of Superpower rivalry. Since the conflict in Indo-China (Vietnam war as well the Cambodian conflict) became an impediment in the realisation of ZOPFAN, a nuclear weapon-free zone proposal was introduced as an interim measure which could be advanced as a prelude to the realisation of ZOPFAN. Hence, for the understanding of conditions leading to the establishment of SEANWFZ, it is important that it should be discussed as part of ZOPFAN.

The ZOPFAN proposal was put forward in 1971 in Kuala Lumpur Declaration by ASEAN. The purpose was to bring about a relaxation of tension and lasting peace in Southeast Asia through the prevention of intervention by outside powers and encouragement of greater cooperation among the indigenous states.[2]

Owing to its geostrategic position due to vital sea lanes such as Malacca, Ombai, Welar, Sunda, etc., between the Indian and the Pacific oceans and between Asia and Australia, Southeast Asia got exposed to outside intervention for a long time. Moreover, since World War II, the region became an arena of internecine warfare due to involvement of the great powers in the Indo-China conflict. As a matter of fact, intervention of outside powers was accelerated as a response to the needs of regional powers. The conflicting global interests of the two Superpowers seemed to have been epitomised, at least for some time in the ideologically divided Southeast Asia.[3] The region eventually became the centre for one of the longest and fiercest ideological struggles: "nationalism vs colonialism/imperialism, socialism/communism vs liberalism/capitalism and religious strife."[4]

Thus, the threats to peace and stability in Southeast Asia were both at extra-regional level as well as at intra-regional level. At the extra-regional level, the threat was with regard to intervention by outside powers. The major powers active in the region were the US, USSR, China and to a lesser extent Japan. And at the intra-regional level, the threat was due to the existing disputes among regional countries.

EXTRA-REGIONAL THREATS

For Southeast Asia the relief which came with the end of World War II was accompanied by the anxiety of the beginning of the Indo-china war. In fact, the conflict in Indo-China is considered as the prime reason for the intensification of the outside intervention in the region. In order to understand this conflict in totality it is important to have a glance at the problems which existed in Indo-China.

After 1945, the French, with the help of the British, had established French rule which stood against the communist forces in South Vietnam while the northern part was under communist influence ruled by Viet-Minh Party. However, while the communist victories in neighbouring China resulted in the establishment of closer links with northern Vietnam, USA responded, by offering aid to the French side. It is at this point that the war assumed a cold war dimension. The Geneva Conference of 1954 in this regard could hardly bring in any meaningful relief. Although a ceasefire was signed yet, in effect, Vietnam was divided along the 17th parallel and thus, two governments were recognised—Democratic Republic of Vietnam with its capital in Hanoi in the north and State of Vietnam in the south with Saigon as its capital under the French rule. The partition of Vietnam brought about severe economic and social disruptions on both the sides and greatly expanded influence for the major benefactors of each side—the United States for the south and erstwhile Soviet Union and China for the north.[5] Very soon South Vietnam moved totally under the US sphere of influence depending heavily on US aid. The US forces were also sent to provide military assistance to South Vietnam in its war against Communist forces. The US policy towards Vietnam was basically governed by the doctrine of 'Containment of Communism'. Pursuing this policy, the then US Secretary of State, John Foster Dulles put forward the 'united action' approach which resulted in the 'Manila Pact' in 1954. The Manila Pact, in turn created the Southeast Asian

Treaty Organisation (SEATO), its members being the US, the United Kingdom, France, Australia, New Zealand and among Asian powers, Pakistan, Thailand and the Philippines. As a matter of fact, SEATO represented extention of Cold War in Southeast Asia.[6] The major Non-Aligned powers—India, Indonesia and Burma—opposed the pact. Over the issue of Indo-China conflict also the regional countries on the whole were not united. Laos and Cambodia maintained official neutrality towards both the Vietnams, but held little optimism that the South would emerge victorious. The Non-Aligned states avoided direct interaction with South Vietnam. While the countries like Malaysia and Singapore gave verbal support to non-communist South, the Philippines and Thailand were not only parties to SEATO but also provided military bases to the US and were directly involved in the war.[7]

Meanwhile, in the US with the death of Kennedy, President Johnson came to power with a commitment to defend South Vietnam at all costs. By 1968, the US forces in South Vietnam were measuring more than 500,000 along with contingents from Australia, the Republic of Korea, the Philippines and Thailand. In turn, the communist forces under North Vietnam became increasingly dependent on China and USSR for aid and support.[8] Thus, one can see that the war in Indo-China became an extension of the Cold War with heavy involvement of the US, USSR and China and all of them being nuclear powers the nuclear threat was apparent. Also, there was a heavy movement of nuclear-powered ships through the sea-lanes and military bases were present in the region. While the US maintained two military bases in the Philippines—Subic Bay and Clark air base—former Soviet Union had two bases in Vietnam Cam Ranh Bay and Danang Air Base. The two Superpowers always followed a policy of not disclosing the presence of nuclear weapons at these bases. Besides, on several occasions the situation came on the verge of involving nuclear weapons. For example, during the battle of Dien Bien Phu in 1954, between the French army and the Viet Minh forces the US Secretary of State, John Foster Dulles, thought of using nuclear weapons to save the beleaguered French army from total extermination by Viet Minh. But since the action had to be jointly taken with Britain it could not materialise. The British government refused to go along with the US stand for fear of a third world war.[9] During clashes between Soviet Union and China over Usuri river in 1969 also there was a fear of extension of the war to Southeast Asia as well. Due to these nuclear threat loomed large over Southeast Asia.[10]

REGIONAL THREATS

Apart from the threats due to involvement of external powers, the region also contained in itself various seeds of potential conflict. At the regional level there existed several problems in Southeast Asia during the 1950s and the 1960s. Other than the economic problems and fragility of political structures, the newly independent countries of Southeast Asia were involved in several bilateral disputes. Indonesia and the Philippines were opposed to the formation of Malaysia. In 1962, the Philippines claimed the territory of North Borneo (Sabah) in opposition to British plan to make it a part of Malaysia. Indonesian President Sukarno followed a policy of 'confrontation' (Konfrontasi) towards his neighbours, especially Malaysia.[11] There was a dispute for claims over Sabah which involved Malaysia, Indonesia as well as the Philippines.[12] The island state of Singapore was also at the centre of such tensions which arose from its separation from Malaysia and the government of Singapore exhibited an acute sense of vulnerability.[13]

However, despite these bilateral problems, the desire for regional cooperation was deep-rooted in the Southeast Asian countries and regionalism was seen as the first step towards regional security. To quote Jusuf Wanandi, "Regionalism in Asia-Pacific is critical to maintaining regional peace and security. Only on the basis of this regionalism can common goals and objectives be established, Confidence Building Measures (CBMs) be nurtured and trust be built in the region."[14]

One of the first initiatives in this direction was taken by Prime Minister Tunku Abdul Rahman of Malaysia and the Association of Southeast Asia (ASA) was formed in 1961 with Thailand, Malaysia and the Philippines being the founder-members. But the association was rendered useless due to territorial dispute between Malaysia and the Philippines over Sabah. Another such initiative came in the form of MAPHILINDO in 1963, an initiative of Philippines President Diosdado Macapagal. The association, however, became irrelevant before it could properly take off due to 'Crush Malaysia' policy launched by President Sukarno.

Nevertheless, in 1967, Association of Southeast Asian Nations (ASEAN) was successfully formed and till date is one of the most important regional organisations. The formation of ASEAN itself can be regarded as a major step towards regional security arrangement. Though political considerations were played down upon, there are good reasons to believe that it was of primary importance.[15] The Bangkok Declaration clearly proclaimed its goal as, "to promote regional peace and stability through abiding respect for justice and the rule of law in the relationship among the countries of the region and adherence to the principles of United Nations Charter."[16] It boosted the stability and cooperation in the region.

Thus looking at the general atmosphere of Southeast Asia during the post-World War II era, it can be understood as to why a need for a much more broader concept with regard to regional security, like ZOPFAN was felt with NWFZ as only one of its components. This is the reason why the proposal had two levels of implementation. At the first level Southeast Asian countries were expected to espouse non-aggression and devise ways and means to ensure peace and security among themselves, and at the second level external powers were expected to be refrained from making Southeast Asia an arena for their conflict and power ambitions. In this regard, however, the question arises why this particular time, i.e., 1971, was chosen by the ASEAN countries to adopt the concept of ZOPFAN. In fact, in vast areas such as Asia-Pacific, regional confidence building, security and disarmament issues cannot exist alone: on the contrary, they are inevitably closely linked to and influenced by what is happening in other parts of the world.[17] Likewise, during that period there were certain developments at regional and international levels which prompted the declaration ofZOPFAN.

At the international level some of the major developments which took place are as follows.

BRITISH WITHDRAWAL FROM EAST OF SUEZ

The British military presence in Singapore and mainland Malaysia was mainly guided by the protection of trading routes to its dominions in the area and their internal security.[18] In September 1954, Britain became a member of Southeast Asia Treaty Organisation (SEATO) in order to expand its role in the security arrangements of the area. In 1957, it signed the Anglo-Malayan Defence Agreement with the independent Malaysian government which was again expanded in 1963.[19] But after 1966, the situation changed. A strong economic constraint was felt, and in 1968 with the devaluation of Sterling there came a reversal of policies. The Wilson Administration announced its decision to withdraw all military forces and facilities from Singapore/Malaysia by the mid-1970s. At first, the announcement was received with good grace since there was sufficient time to make alternative arrangements and also there was a provision for compensation.[20] But the announcement was revised after a few months and it was stated that the withdrawal was to be completed by 1971. Although Britain tried to give assurance of its continued participation in the security of the area in the form of Five Power Defence Agreement, yet Malaysia and Singapore expressed concern over the announcement. They 'stressed the importance which they attached to the defence agreements with Britain'.[21]

In terms of the consequences of the British withdrawal there were two major concerns—economic and strategic. In economic terms, the consequences were to be dire.[22] In terms of strategic environment of the area there arose a fear of creation of a power vacuum. This became a matter of grave concern for the ASEAN countries as the situation could have led to the augmentation of the great power rivalry in the region.[23] Subsequently, the proposal for neutralisation was put forward by Malaysia in 1970.

THE GUAM DECLARATION

After the elections in the US in 1968, Richard Nixon became the President of America with a pledge that he would terminate the war in Vietnam. In this regard he laid down an outline of a new foreign policy for America in a press conference held in Guam in July 1969 which came to be known as the "Guam Declaration".[24]

The blueprint of Nixon's policy was presented as "US Foreign Policy for the 1970s: A Report to the Congress by Richard Nixon". It is also called the Nixon Doctrine. Vide this Doctrine it was clear that Americans would be with drawing from Indo-China but President Nixon also tried to convince the allies while stating: "abrupt shifts in our policies—no matter how sound in concept, are unsettling", and that "precipitate shrinking of the American role would not bring peace". Thus he assured the ASEAN leaders, that "US will keep all its treaty commitments".[25]

The nuclear issue also figured in the Nixon Doctrine which held that: "US will provide nuclear shield, if a nation allied with it, or if a nation whose survival the United States considers vital to its security and the security of the region as a whole."[26]

However despite these assurances, the Nixon Doctrine represented a shift in the United States strategy in Southeast Asia which created a panic among ASEAN members. Besides, the growing tilt of the United States towards China was also a matter of concern for the countries of Southeast Asia.

The incident which had actually caught the attention of the ASEAN countries was the announcement made in July 1971 by President Nixon after the secret visit of Henry Kissinger to

Peking, about his forthcoming visit to China. This radical shift in the US foreign policy was brought about without prior consultations with or information to its close allies. Obviously, ASEAN members, particularly the Philippines and Thailand, felt betrayed and nervous. These countries began to reassess their policy in the light of new developments, and the need for a new approach towards regional security was felt.[27]

The new American policy was prompted by the changing equations due to a split in the Sino-Soviet relations. In fact, signs of rift between China and USSR had surfaced as early as 1959 itself. A decade later, by 1969, China and USSR got engaged in bloody clashes over their borders along Usuri river. By the 1970s, the Sino-Soviet split had become an undeniable fact and had set a pattern for forging new relationships among nations all over the globe. The Kuala Lumpur declaration indicates the intent of shifting form traditional pro-US, anti-communist policies to one of seeking a contact and dialogue with the communist countries.[28]

By the end of the sixties, it had become clear that the Indo-China war was un-winnable for Americans. This made the regional countries like Malaysia to look for other security options and when the statement from the Guam Declaration of President Nixon was made that the US would withdraw from Indo-China and that the regional allies would have to take care of their security by themselves with the help of United States,[29] the need for a new regional order was urgently felt.

GROWING SOVIET INTEREST

Southeast Asia did not contain an interest considered vital for Moscow until the mid-1960s. Eurocentricism, lack of substantial ties and narrow and rigid policies followed by Stalin had led to a low Soviet involvement in the region. In fact, the influence of China, the United States and Japan, was of more concern for Soviet Union rather than the region itself.[30]

After World War II, however, the Soviet Union established diplomatic relations with Thailand in 1946, with Burma in 1948 and with Indonesia in 1950. It also recognised the Democratic Republic of Vietnam in 1950. Several developments in late 1960s such as British withdrawal from East of Suez, American withdrawal from mainland Southeast Asia and the intensification of Sino-Soviet conflict led the Soviet Union to forge closer ties with Asian countries. After the death of Stalin, when Nikita Khrushchev came to power, the Soviet foreign policy underwent quite a radical change, the basis of which was to forge closer ties with the Third World countries, especially with those who were critical of the United States and its allies.[31] Hence, after 1969, the Soviet Union had become increasingly active in the region. The increased inability of the Soviet navy and merchant fleet, particularly with the granting of berthing facilities by Singapore in 1970 for Soviet use, was at least in part intended to present a strategic, psychological and economic counterweight to the United States.[32]

The 'Asian Collective Security' proposal, put forward by Leonid Brezhnev in 1969, was also a part of the same strategy. The idea was put forward in his address to the World Conference of Communist and Workers Party held at Moscow. He said: "Despite the pressing problem of the present international situation, we do not push into the background more long range tasks, especially the creation of a system of collective security in those parts of the world where the threat of the unleashing of the new world is centered."[33] It was a vague scheme but, in essence, consisted of 'friendship treaties' with potential Soviet allies and those predisposed towards the Soviet Union.

However, Soviet diplomatic overtures in the region did not achieve the desired objectives and the ASEAN countries remained suspicious about Soviet initiatives. But the Soviet endorsement of ZOPFAN proposal did facilitate in forging its relations with ASEAN countries.[34] By late 1970s, the Soviet Union had started giving uncritical support to Vietnam in its quarrels with Khmer Rouge. The increasing operations of its

Pacific fleet and military presence in Cam Rahn and Danang in Vietnam also testifies Soviet Union's growing involvement in the region.

The ASEAN countries on the other hand were concerned about the negative Superpower rivalry. In their effort to keep away the external powers, they realised the relevance of ZOPFAN and NWFZ.

THE COMMUNIST THREAT

The communist threat remained one of the major factors in the politics of Southeast Asia till the end of the Cold War. The American policies in the region were shaped in reference to its interests vis-a-vis the Soviet Union and China.[35] The American decision to defend South Vietnam and Thailand, contain spread of communism in Indo-Chinese peninsula, form military bases in the Philippines and Thailand and the formation of SEATO, all these steps were motivated by the policy of 'containment of communism'. Britain also planned to use commonwealth to build up regional cooperation against communism. The communist victories in China in 1949 caused a major alarm in the Western bloc as they feared of a 'domino effect', meaning that the triumph of communism in one country would destabilise its neighbour, each falling in turn to communist forces.[36]

The regional countries also, in their post-independence period, operated as a sub-system that reflected the bi-polar nature of contemporary international relations. Generally speaking, all the maritime Southeast Asia and Thailand adopted a pro-Western and anti-communist outlook in their domestic and foreign policies.[37] However, almost every Southeast Asian country faced communist-led subversive movements which attracted strong following among some patriots, minorities and the poor. To many, communism was the hope promising a fairer society with freedom from foreign domination, land for the peasants, and a better standard of living. Communist risings took place in Burma, Malaya and Indonesia in 1948. Indonesia faced an abortive coup in 1965 supported by the Indonesian Communist Party (PKI) against the Sukarno government. The communist insurgencies in the Federation of Malaya in 1948 resulted in the declaration of State of Emergency which lasted for 12 years. In Indo-China, the national struggle for independence became intertwined with the revolutionary ideology of communism. Thailand, following the communist victory in China, developed a strong anti-communist position as the Thai Communist Party at that time was dominated by the ethnic Chinese. The Philippines also suffered from Sino-phobia. The movements were generally supported by China and to a lesser extent, by Soviet Union. China maintained party to party contact with the Communist parties of Southeast Asia.[38] Meanwhile, the US taking its policy of 'containment of communism' a step further, instigated a coup in Cambodia in March 1970 in which Prince Sihanouk was replaced by General Lon Nol.

By extending Vietnam war to Cambodia, the Americans strengthened the Communist forces of the area.[39] Prince Sihanouk, after the coup, commented that "the Socialist camp considers recent events are a direct threat to the balance of forces between the United States and itself."[40] On the other hand, there was an appeal from Cambodia to Indonesia for military assistance. In April 1970, it was reported that an Indonesian military mission was in Phnom Penh to consider Cambodia's appeal.[41] These developments of getting directly involved into the affairs of Indo-China certainly alarmed the ASEAN countries.

Thus, the most important motivation behind the idea of neutralisation of Southeast Asia seemed to be to keep away all the external powers from interfering in the internal affairs of the region. Speaking before a meeting of the Press Foundation of Asia Assembly in September 1971, Adam Malik, the foreign minister of Indonesia, urged that ". . . . the nations of Southeast Asia should consciously work towards the day when the security of their own region will be the primary responsibility of

Southeast Asian Nations themselves.[42] However, for the first time, the idea of neutralisation was put forward by Malaysia.

As a matter of fact, for Malaysia, the withdrawal of the US coupled with reduced British role (which had retained the responsibility of Malaysian defence since its independence) was a situation of great alarm. Such a situation would have left the region potentially vulnerable to the People's Republic of China and USSR. Malaysia was particularly afraid of China which supported the Malaysian Communist Party (MCP) in the 1947-57 emergency.[43] Thus it was Malaysia's own security concern which made it to put forward the idea of neutralisation for Southeast Asia.

However, any definite plan could not take shape immediately. The first definite step in this direction was the formation of 'Peace Plan' in 1968 by Malaysia's Tun Dr. Ismail Abdul Rahman wherein he proposed the idea of neutralisation for Southeast Asia.[44] Dr. Ismail later on became Deputy Prime Minister in Tun Abdul Razak's government, who adopted the essential ideas of the Peace Plan and converted it into a proposal for a Zone of Peace, Freedom and Neutrality. Later on this proposal was elaborated by the government's theoretician, Tan Sri Ghazali Bin Shafie.

MALAYSIAN PROPOSAL OF NEUTRALISATION

Elaborating the concept of neutralisation Mr. Shafie stated, "Southeast Asia is a factor in the calculations of the major powers. It is natural to expect that their interest in the region will grow. . . . All these will add to potentially dangerous situations, unless clear ground rules are established in Southeast Asia. In the interest of the peace in area, and very likely of the world, the entire Southeast Asian region should be neutralised. The countries in the region should get together, and as a group, serve notice to the world that the region should no longer serve as a theatre for international power struggle."[45]

The Malaysian proposal for neutralisation had two important aspects and can be viewed on two levels. At the first level the countries of Southeast Asia should agree upon:

"Individual countries in the region must respect one another's sovereignty and territorial integrity, and not participate in activities likely to directly or indirectly threaten the security of another. This is an essential requirement. Non-interference and non-aggression are basic principles which Southeast Asian countries must unequivocally accept before any further steps can be taken." Moreover,

i) All foreign powers should be excluded from the region.
ii) The region should not be allowed to be used as a theatre of conflict in the international power struggle.
iii) They should devise ways and means of and undertake the responsibility for ensuring peace among member states.
iv) They should present a collective view before the major powers on vital issues of security.
v) They should promote regional cooperation.[46]

At the second level, guarantee was taken from major powers, i.e., US., USSR and China. Dr. Shafie further explains:

i) Southeast Asia should be an area of neutrality.
ii) The powers undertake to exclude countries in the region from the power struggle among themselves.
iii) The powers should devise the supervisory means of guaranteeing Southeast Asia's neutrality in the international power struggle.[47]

Thus, according to this plan the regional countries were to be responsible for maintaining peace among themselves whereas the external powers had to give guarantee to prevent conflicts which are externally inspired. In a speech in 1971, Ghazali Bin Shafie said: ". . . . it is Malaysia's hope that nonaligned countries will be able to endorse the neutralisation not only of Indo-China but of the entire region, guaranteed by three major powers, the Peoples Republic of China, the Soviet Union and the United States against any form of external interference, threat or pressure".[48] Thus, the Malaysian proposal can be seen as part of the "tactical response to the need for a defensive political structure in the area. Moreover, the proposal further coincided with the emergence of a mutual or shared interest among the major powers to quarantine volatile region which could propel them unwittingly into self-destructive, or certainly unproductive, enterprises."[49]

Initially, in an effort to gain support, Malaysia put forward the proposal of neutralisation at fora like Lusaka Summit of Non-Aligned Nations in 1970; the commemorative session of the 25th anniversary of the United Nations in 1970; and the Commonwealth Conference in Singapore in 1971. By 1974, Malaysia had established diplomatic relations with the People's Republic of China and also won its support for ZOPFAN.[50] However, the United States remained silent[51] and Soviet Union did not favour it.[52]

As far as the regional powers are concerned most of them viewed the idea as impractical and doubted its utility in providing for their security. The Indonesian response was cool but not negative. It saw some merit in the attempt to regulate the activities of external powers in the region. However, the initial Thai reaction was hostile as it viewed its security relations with the United States as critical, whereas Singapore and the Philippines were sceptical about the proposal.[53]

In spite of this scepticism, there was a widespread belief in the whole of Southeast Asia that the area had suffered a lot because of the rivalry of external powers. This was the reason why ZOPFAN has always held some ground and in fact was never a rejected idea. But at the same time the regional powers also found it important to maintain the existing security arrangements and balance of power. The Kuala Lumpur Declaration which followed the Malaysian proposal had many modifications on what Malaysia had originally proposed. The ZOPFAN proposal was endorsed by the ASEAN states at Kuala Lumpur in 1971.

REFERENCES

1. Acharya, Amitav and Sola Ogunbanavo, "The Nuclear Weapon-Free Zones in Southeast Asia and Africa", *SIPRI Yearbook of Armament Disarmament and International Peace* (Oxford University Press, 1998), pp.143-144.
2. Alagappa, Muthiah, "Regional Arrangements and International Security in Southeast Asia: Going Beyond ZOPFAN", *Contemporary SoutheatAsia*, Vol.12, No.4, March 1991.
3. Djiwandono, Seodjati, "Cooperative Security in the Asia Pacific Region: An ASEAN Perspective", *The Indonesian Quarterly*, Vol.22, No.3, Third Quarter, 1994, p.207.
4. Habib, Hasnan, "ASEAN in Search of Peace and Stability in Southeast Asia", *United Nations Disarmament Workshop for Asia Pacific* (New York: United Nations, 1991).
5. McCloud, Donald G., *System and Process in Southeast Asia: The Evolution of a Region* (Colorado: Westview Press, 1986), p.198.
6. Sardesai D.R., *Southeast Asia: Past and Present* (New Delhi: Vikas Pub., 1981), p.400.
7. McCloud, pop.cit, pp.198-99.
8. See for example, Kaiser, David, *American Tragedy: Kennedy, Johnson and the Origin of Vietnam War* (Massachusetts: Harvard University Press, 2000).
9. D.R. Sardesai, op.cit., p.195.
10. Dibb, Paul, "The Interests of Soviet Union in Southeast Asia and Southwest Pacific: Implications for Regional Security", Paper presented at the *Conference on International Security in Southeast Asia and Southwest Pacific Region*, SDSC, ANU, Canberra, 12-15 July 1982.
11. Gordon, Bernard K., *The Dimension of Conflict in Southeast Asia* (Englewood Cliffs, N.J.: Prentice Hall, 1966), pp.9-40.
12. Seodjati Djiwandono J., "Intra-ASEAN Territorial Disputes: The Sabah Claims", Paper presented at a seminar on *ASEAN into 21st Century: Dealing with Unresolved Issues*, organised by ASEAN-ISIS in Manila, 14-15 January 1994.
13. Samad, Paridah Abdul, and Muhammad Mokhtar, "ASEAN's Role and Development as a Security Community", *Indonesian Quarterly*, Vol.23, No.1, 1995, p.71.

14. Wanandi, Jusuf, "Regionalism in the Asia Pacific Region", *The Indonesian Quarterly*, Vol.28, No.2, 2000, p.226.
15. Naidu, G.V.C., "Neutralisation of Southeast Asia: Zone of Peace, Freedom and Neutrality", *Strategic Analysis*, Vol.10, No.3, June 1986, p.57.
16. Text of ASEAN Declaration.
17. Huasun, Qui, "An Approach to Confidence Building in Asia Pacific Region", *Topical Paper* 6 (New York: United Nations, 1991), p.78.
18. Leifer, Michael, (ed.), *Constraints and Adjustments in British Foreign Policy* (London: George Allen & Unwin Ltd., 1972), pp.86-87.
19. Ibid., p.87.
20. Ibid., p.89.
21. *The Times*, 9 January 1968, p.1.
22. Ibid., 10 January 1968, p.23.
23. Ibid., 9 January 1968, p.1.
24. For details, see Guam Declaration, published in *Southeast Asia under New Balance of Power* (New York: Praeger Publishers, 1974), pp.122-9.
25. Taken from the Text of the Nixon Doctrine, published *in Southeast Asia under New Balance of Power* (New York: Praeger Publishers, 1974).
26. Ibid.
27. K.P. Saxena, *Cooperation in Development: Problems and Prospects for India and ASEAN* (New Delhi: Sage Publication), p.31.
28. Ibid., p.31.
29. Naidu, G.V.C., "Neutralisation of Southeast Asia: Zone of Peace, Freedom and Neutrality", *Strategic Analysis*, Vol.10, No.3, June 1986, p.358.
30. Horn, Robert C., "The Soviet Perspective", *Southeast Asia Under New Balance of Power* (New York: Praeger Publishers, 1974), p.30.
31. Alagappa, Muthia, "Major Powers and Southeast Asia", *International Journal*, Vol.44, No.3, September 1989, pp.553-4.
32. Horn, n.30, p.47.
33. Chairman Brezhnev Reference to Collective Security, reported in *Pravda*, 8 June 1969.
34. For details see, Noorani, A.G., *Brezhnev Plan for Asian Security: Russia in Asia* (Bombay, 1975).
35. Simon, Sheldon W., "United States Security Policy and ASEAN", *Current History*, Vol.89, March 1990, p.98.
36. For the effects of the cold war and communist insurgencies in Southeast Asia, see Nicholas Tarling, *The Cambridge History of*

Southeast Asia, Vol.II (Cambridge: University Press, 1992), pp.596-606.

37. Ibid., pp-596-606
38. Ibid., pp-596-606.
39. Naidu, G.V.C., "The Kampuchean Problem: Problems and Prospects", *Strategic Analysis*, January 1987, p.1232.
40. *The Times*, 19 March 1970.
41. Soon, Lau Teik, Indonesia and the Regional Security", *The Djakarta Conference on Cambodia* (Singapore, 1972), p.6.
42. *Far Eastern Economic Review*, 25 September 1971, p.31-33.
43. Palmer, Ronals, G., and Rockford, Thomas J., *Building ASEAN: 20 Years of Southeast Asian Cooperation* (Washington: Centre for Strategic and International Studies, 1987), p.12.
44. For details, see Saravanamuttu, J., *The Dilemma of Independence, Two decades of Malaysia's Foreign Policy, 1967-1977* (Penang: Penerbit University Sains Malaysia, 1983), pp.74-75.
45. M. Ghazali bin Shafie, "The Neutralisation of Southeast Asia", *Pacific Community*, Vol.2, No.2, 1971, p.114.
46. See M. Ghazali bin Shafie, "The Neutralisation of Southeast Asia", *Foreign Affairs* (Malaysia), No.4, 1971.
47. Ibid.
48. Cited in Foreign*Affairs Malaysia*, Vol.30, 1970, p.37.
49. Clark, Claire, ed., *Australian Foreign Policy: Towards a Reassessment* (Cassell Australia, 1973), p.123.
50. Saravanmuttu, J., "ASEAN Security for 1980s: the Case for a Revitalised ZOPFAN", *Contemporary Southeast Asia*, Vol.6, No.6, September 1984, p.187.
51. For details see, *Neutralisation of Southeast Asia: Problems and Prospects* (Washington: U.S. Government Printing Office, 1976).
52. In 1968 President Brezhnev proposed quite a vague solution to the problem for neutrality for the region in the form of Asian collective security system. For details, see Sheldon W. Simon, "The Soviet Union and Southeast Asia: Interests, Goals and Constraints", *Orbis*, Vol.25, No.1, 1981, pp.
53. Alagappa, op. cit, p.273.

6

In Search for a Regional Security Framework for Southeast Asia: ZOPFAN, TAC and NWFZ

THE KUALA LUMPUR DECLARATION

The Kuala Lumpur declaration was issued by the foreign ministers of Indonesia Malaysia, the Philippines, Singapore and special envoy of the National Executive council of Thailand.

In the very beginning of declaration only the two basic motives, i.e., regional cooperation and relaxing international tension, have been described which are follows :[1]

> Firmly believing in the merits of regional cooperation which had drawn our countries to cooperate together in the economic, social and cultural fields in the association of Southeast Asian Nations.
>
> Desirous of bringing about a relaxation of international tension and of achieving a lasting peace in Southeast Asia.

The motives expressed in the above two statements can also be seen in the later part of the declaration which reads:

> Recognising the right of every state, large or small, to lead its national existence free from outside interference in its internal affairs as this interference will adversely affect its freedom, independence and integrity.

> Dedicated to the maintenance of peace, freedom and independence unimpaired.

Apart from these two motives, the declaration was also inspired by the aim of international peace and stability, which is as follows:

> Inspired by the worthy aims and objectives of the United Nations, in particular by the principles of respect for the sovereignty and territorial integrity of states, abstention from the threat or use of force, peaceful settlement of international disputes, equal rights and self determination and non-interference in the internal affairs of states.

This desire can also be seen in the following sentence:

> Believing in the need to meet recent and new developments by cooperating with all peace and freedom loving nations, both within and outside the region, in the furtherance of world peace stability, and harmony.

The Declaration also expressed its belief in the Bandung Conference of 1955. It says:

> Believing in the continuing validity of the Declaration on the Promotion of World Peace and Co-operation of the Bandung Conference of 1955, which among others, enunciates the principles by which states may co-exist peacefully.

Incidentally, the Bandung Conference was held in Bandung (Indonesia) in April 1955 in which 25 Afro-Asian countries participated and the main issues of the conference were economic and cultural cooperation, human rights and self determination, problems of dependent peoples and world peace.[2] Regarding world peace and cooperation a declaration was issued in which there were 10 principles on peaceful co-existence. The Asian African Conference declared its conviction that friendly cooperation in accordance with those principles would effectively contribute to the maintenance and promotion of international peace and security, whilst cooperation in the economic, social and cultural fields would help bring about the common prosperity and well being of all.[3] The Kuala Lumpur Declaration reiterated faith in the 'Declaration on the promotion of world peace and cooperation of the Bandung Conference.

The Kuala Lumpur Declaration also reiterated its commitment to the principles of the Bangkok Declaration which established ASEAN in 1967, "that the countries of Southeast Asia share a primary responsibility for strengthening the economic and social stability of the region and ensuring their peaceful and progressive national development, and they are determined to ensure their stability and security from external interference in any form of manifestation in accordance with the ideals and aspirations of their people".

The Kuala Lumpur Declaration had also put forward the idea of establishing nuclear-free zone in Southeast Asia, on the line of the treaty for the Prohibition of Nuclear Weapons in Latin America and Lusaka Declaration proclaiming Africa a nuclear-free zone, for the purpose of promoting world peace and security by reducing the area of international conflicts and tensions.

Finally, after the above mentioned preamble the declaration was made stating:

> That Indonesia, Malaysia, the Philippines, Singapore and Thailand are determined to exert initial necessary efforts to secure the recognition of, and respect for Southeast Asia as a zone of peace, freedom and neutrally, free from any form or manner of interference by outside powers.
>
> That Southeast Asian countries should make concerted efforts to broaden the areas of cooperation which would contribute to their strength, solidarity and closer relationship.

As is clear from the forgoing discussion that stopping external interference and regional cooperation were the two basic motives behind the Kuala Lumpur Declaration. Also, in accordance with the concept of the Zone of Peace as discussed in Chapter 1, ZOPFAN had to be implemented at two levels—at level one, the region was to be prevented from becoming an arena for the rivalry of external powers and at the next stage collective self-reliance was sought to be achieved through peaceful settlement of disputes, peaceful co-existence, etc.

To quote Hasnan Habib, ZOPFAN should be seen as a "set of code of conduct between ASEAN state themselves, between ASEAN states and outside powers, and among outside powers themselves, within the context of maintaining peace, security and stability in the region."[4]

In this regard, it will be important to discuss the attitudes of external as well as regional powers towards the proposal of ZOPFAN.

Reaction of External Powers

As far as United States is concerned it was deeply involved in Southeast Asia. The US involvement apart from strategic and economic, was, as discussed earlier, basically born out of attempts to contain communism. This had resulted in American military presence in many more states than Vietnam such as Thailand, the Philippines, Taiwan, South Korea, etc. However, the idea of neutrality was not new to the US. There was a proposal for creating neutral zones in Asia as a part of policy of containment and a study was conducted in this regard for the Committee on Foreign Relations of the United States Senate in 1966.[5]

Nevertheless, the United States was opposed to the idea of ZOPFAN keeping in view the existing balance of power and above all the existing security arrangement (multilateral like SEATO as well as bilateral such as with the Philippines and Thailand) and also the US bases in the Philippines and Thailand.[6] Moreover, it was important for the US to safeguard pro-US regimes in Cambodia and Laos from falling to communists.

As far as erstwhile USSR was concerned its main interest in Southeast Asia during the 1960s and the 1970s remained containment of USA and China. The Indo-China war led to a greater Soviet involvement which was further accelerated after the American instigated coup in Cambodia in 1970. Nevertheless, Soviet Union supported the ZOPFAN proposal as neutral Southeast Asia was preferable to a pro-Western one. Soviet support would have also given it a legitimate role in the region. There was even an attempt to incorporate ZOPFAN into 'Asian Collective Security' proposal by Brezhnev. However, the uncritical support given to Vietnam against Khmer Rouge and China and acquisition of military facilities in Cam Ranh Bay and Danang, in fact, contradicted Soviet support for ZOPFAN.[7]

The people's Republic of China, keeping in view its geographical proximity with Southeast Asia, has always been in the most advantageous position among the external powers active in the region. There are several dimensions of Chinese policy towards the region such as political, diplomatic, strategic, economic, ideological and so on. As far as its attitude towards ZOPFAN is concerned, although Beijing took some time but by mid-1970s it endorsed the proposal.

As already discussed in Chapter 1 China had been supportive of such proposals in the United Nations. However, the endorsement of ZOPFAN probably was motivated by Beijing's efforts to woo ASEAN as its support was seen valuable to counter Vietnamese and Soviet ambitions in Southeast Asia. Secondly, it was realised that if the other powers are excluded from the region China due to its geographical proximity will remain the only influential power in Southeast Asia and finally, it would prevent hostile coalition on its southern flank.[8] However, China did put some conditions such as abolition of bilateral and multilateral military alliances as well as foreign bases from the area, before endorsing the proposal.

Japan, another external power which was beginning to play an important role in Southeast Asia under the US umbrella, remained rather non-committal towards the issue. On the one hand, it had to see the US interests, and on the other hand, its own peace constitution, anti-nuclear policies and sensitivity to ASEAN concerns was also important.[9]

Reaction of Regional Powers

As far as realisation ofZOPFAN is concerned, differences existed among the ASEAN countries also. The Declaration was issued only after considerable wrangling among the ASEAN members, but the different legal and policy implications of 'neutrality' remained in contention. These differences however were largely in terms of nuance rather than substance and can be explained as a reflection of different historical backgrounds, geopolitical factors, internal problems and the perception of threat in the individual countries. To some defence needs were to be met individually, to some bilaterally and for others collectively.[10]

As we have seen Indonesia and Malaysia were the countries which had supported the idea of neutralisation of Southeast Asia and had been the main force behind SEANWFZ. They had also signed friendship treaties between them but had different security concerns. As regards Malaysia, it may be noted that it still had stormy relationship with Singapore over its separation from Malaysian confederation in 1965. However, in the very same year, that is, 1971, Malaysia entered into Five Power Defence Arrangement with Singapore, the United Kingdom, Australia and New Zealand which provided that "...in the event of any form of armed attack externally organised or supported or that of such an attack against Malaysia and Singapore, their governments would immediately consult together for the purpose of deciding what measures should be taken jointly or separately in relation to such attack or threat."[11]

Singapore favoured in maintaining balance of power in the region. It also supported the American bases in Philippines and was against ridding the region of friendly external powers. A disturbed region would have also affected Singapore's business prospects. It had even proposed for a permanent naval task force of Japan, the US, Australia and other European powers to patrol Southeast Asian waters,[12] hence Singapore's reaction was cautious as it was not keen on the idea. Singapore's foreign minister, S. Rajarathan, after Kuala Lumpur meeting said, "We may or may not be able to persuade the big powers to accept the neutral location."[13]

Thailand, however, was probably the only country where the neutralisation plan received the least support. It had worries related to prospective withdrawal of American forces from Vietnam and was not prepared to close down American bases. It was bound by defence treaties with the US. Moreover, the country was facing communist aggression and being a frontline state was also worried about Soviet and Vietnamese aggressions. The Deputy Chairman of Thailand's National Council, General Prapas Charustian, was reported to have expressed disapproval for the neutralisation plan.[14]

The Philippines took the idea enthusiastically although it depended on the ongoing negotiations over future of American military presence in the region. The US bases were important for the Philippines economically as well as for its defence and it was also bound by the defence treaty with the US. However, the Filipinos hated the US policy of non-disclosure about the presence of nuclear weapons on the US bases.[15]

Since the proposal for ZOPFAN intended to cover the whole of Southeast Asia, including non-ASEAN countries it is important to note their reaction also. As regards Vietnam, although Soviet presence was vital for it, it did not completely reject the idea. However, it first wanted total withdrawal of the US forces. At the Colombo Conference of the Non-Aligned Countries, Vietnam blamed the Southeast Asian non-aligned countries that they were dependent on 'neo-colonialism' and were not sincerely 'non-aligned' and 'neutral'.[16] Laos, was not entirely opposed to the idea of neutralisation, keeping in view its own neutral policy. But encouraged by Vietnam, it blocked Malaysian efforts to seek reaffirmation of the Kuala Lumpur

Declaration on the ground that "the Kuala Lumpur Declaration was issued at the same time when the US was intensifying its war of aggression in Indo-China, in order to cover up the participation of ASEAN countries in the US war of aggression."[17] Cambodia, whose situation was a major hurdle in the realisation of ZOPFAN, although depended on China yet along with Burma supported the Kuala Lumpur Declaration.[18]

Thus, although the idea of neutralisation received formal approval of the ASEAN countries but only declaration of such, an idea of regional security was not enough. What was actually required was its effective implementation of such ideas for which reaching consensus among regional countries was most important.

Most of the regional states continued to value the existing multilateral as well as bilateral security arrangements with great powers, keeping in view the uncertain conditions in the region. If the proposal for ZOPFAN had to be pursued, the existing balance of power and security arrangements had to be undermined which would have meant a major shift in the policies of the regional countries.

None-the-less, a not-so-enthusiastic response by the external as well as regional powers did not mean that the concept lost its relevance altogether. As Micheal Leifer explains it, the corporate endorsement of the Zone of Peace gave security ideas a higher profile and correspondingly reinforced ASEAN's claim to prescribe for regional order as indicated in Bangkok Declaration. However, it was devoid of immediate utility other than to serve as a reference point for a symbol of intra-ASEAN solidarity.[19] In the words of David Denoon and others, "vaguely articulated ZOPFAN declaration of 1971 provided principled basis for avoiding position on the Cold War and Indo-China hostilities. It also bolstered ASEAN's credibility in the Third World."[20]

Thus, one may agree that the ZOPFAN proposal might not have been a total success nor was it a total failure. As an extra-ASEAN policy it remained almost at the conceptual level, but the second motive of regional cooperation was equally important and it did make some progress as an intra-ASEAN policy. ASEAN responded to the changes in Indo-China due to American withdrawal by organising its first meeting of heads of the governments in Bali in 1976. In this summit some very important decisions were made such as[21] Declaration of ASEAN Concord and Treaty of Amity and Co-operation.

DECLARATION OF ASEAN CONCORD AND TREATY OF AMITY AND CO-OPERATION

The ASEAN Concord was declared to "consolidate the achievements of ASEAN and expand ASEAN cooperation in the economic, social and political field." It also made reference to the settlement of intra-ASEAN disputes by peaceful means and establishment of an ASEAN sectariat.[22] The Treaty of Amity and Cooperation is a 'code of conduct' for the intra-ASEAN behaviour. In the words of Roeslan Abdulgani, "the Treaty of Amity and Co-operation is the first step taken by ASEAN towards realisation ofZOPFAN."[23]

The Treaty of Amity and Cooperation

The Treaty of Amity and cooperation was signed in 1976 in Bali, with the purpose, to promote perpetual peace, everlasting amity and cooperation among the people which would contribute to their strength, solidarity and closer relationship.[24]

It established the principles to which inter-state behaviour in the region must conform. Article II of the treaty laid down six principles by which relations of the contracting parties were to be governed. Those six principles were:

i) Mutual respect for the independence, sovereignty, equality, territorial integrity, and national identity of all nations.

ii) The right of every state to lead its national existence free from external interference, subversion or coercion.

iii) Non-interference in the affairs of one another.

iv) Settlement of differences or disputes by peaceful means.

v) Renunciation of the threat or use of force.

vi) Effective cooperation among themselves.

Apart from this these six principles, the preamble to the treaty also explains the basic nature of the treaty which states, "Desiring to enhance peace, friendship and mutual cooperation on matters affecting Southeast Asia consistent with spirit and principles of the charter of the United Nations, the ten principles adopted by the Asian-African conference in Bandung on 25 April 1955, the Declaration of the Association of Southeast Asian Nations signed in Bangkok on 8 August 1967 and the Declaration signed in Kuala Lumpur on 27 November 1971."

Thus, the Treaty of Amity and Cooperation can be seen as a genuine effort of the countries of Southeast Asia towards political stability in the region and cooperation among regional countries which was one of the basic motives of ZOPFAN. The Treaty serves a twofold role, first, it establishes a code of conduct governing relations among Southeast Asian Nations—a function which contributes to conflict avoidance; and second, the treaty provides a legal basis for the peaceful settlement of disputes which are consistent with the UN charter. There is no doubt that participation of all the ten Southeast Asian countries has strengthened the treaty immensely.[25] The Bali Summit can be regarded as the high point towards realisation ofZOPFAN.

Along with this summit some other positive developments also took place in the region such as the US withdrawal from the Indo-China, unification of Vietnam and the Paris Peace Agreement which stated "foreign countries shall put an end to all military activities in Cambodia and Laos," and also that the internal affairs of these states "shall be settled by the people of each of these countries without foreign interference."[26]

With this it appeared that a new era of peace and stability would dawn upon the region. There also appeared a new conciliatory attitude of Vietnam towards ASEAN, as there were talks of its possible entry into ASEAN.[27] In 1978, Vietnamese Prime Minister Pham Van Dong toured the ASEAN countries with offers of friendship treaties and presented a counter-proposal for ZOPFAN titled as zone of 'Peace, Independence and Neutrality' from the Vietnamese side.[28]

However, the high morals of communist forces in Vietnam due to American defeat added by a large defence establishment equipped with both Soviet supplies of arms and the arms left behind by the Americans in haste was viewed by the ASEAN countries with great alarm. As a result, national defence budgets of the ASEAN countries began to rise after 1975 and particularly between 1979 and 1980. The high rate of economic growth provided impetus to the process giving rise to a spurt in the regional arms race.[29]

NUCLEAR WEAPON-FREE ZONE IN SOUTHEAST ASIA (SEANWFZ)

The idea of Nuclear Weapon-free Zone in Southeast Asia featured in the preamble to ZOPFAN declaration, which read "cognizant of the significant trend towards establishing nuclear free zone as in the treaty for the prohibition of nuclear weapons in Latin America; and the Lusaka Declaration proclaiming Africa a nuclear-free zone." But ASEAN's interest in this direction was intensified during the early 1980s. Indonesia and Malaysia were the main force behind this proposal. Indonesia has been one of those countries, which attach great importance to Article VII of the NPT, affirming the right of states to conclude regional treaties in order to assure 'total' absence of nuclear weapons in their respective territories. In this regard, Indonesia has always emphasised the regional approach to security.[30] The idea behind the establishment of NWFZ was to provide a platform at which all Southeast Asian countries, then polarised by the Cambodian conflict, could have positive interaction.[31] As a matter of fact, the proposal of SEANWFZ is the first specific proposal for arms control in the region. It can contribute towards the security of zone members by preventing nuclear confrontation and proliferation in the region besides contributing towards the goal of general and complete disarmament.[32]

However, after 1971, when the proposal first came in, it took almost 10 years to begin the work in that direction. The first draft was presented by Indonesia in the Manila Summit. According to the draft, the possession, stationing and manufacture of nuclear weapons by the signatory states was to be banned but the question of transit was ignored. There was also no mention of the military bases of the foreign countries.[33] Whereas, the government of Jakarta and Kula Lumpur wanted to minimise nuclear confrontation, the Philippine government integrated the idea of a zone into talks on the future of US bases.[34]

None-the-less, until the Cambodian problem was solved, no concrete development could take place. Finally, the end of East-West conflict, the settlement of Cambodian problem, the withdrawal of American and Soviet forces from their bases in Southeast Asia and the integration of the three Indo-Chinese countries into ASEAN fold paved the way for signing of the SEANWFZ Treaty. However, the treaty was brought to life as a direct consequence of resumption of the French nuclear testing in the South Pacific in 1995.[35]

The Treaty and Its Provisions

On 15 December 1995, heads of the governments of ASEAN members and its acceding partners signed the Treaty on the Southeast Asia Nuclear Weapon-free Zone. The treaty is not only significant from the regional point of view but it is ASEAN's contribution to the worldwide campaign towards nuclear non-proliferation and disarmament. It is ASEAN's most concrete expression of renunciation of all weapons of mass destruction.[36]

With the declaration of SEANWFZ one more step has been achieved by the countries of Southeast Asia in the direction of realisation of ZOPFAN after the Treaty of Amity and Cooperation. As mentioned earlier, while the implementation of ZOPFAN was kept in obeyance due to problems in Indo-China, SEANWFZ was introduced as an 'stop gap' measure, and was a good example of ASEAN'S pragmatism.[37] In the preamble itself it is stated, "Recalling the Declaration of Zone of Peace Freedom and Neutrality (ZOPFAN)." It is an elaborate treaty with 22 articles and along with seven articles in Protocols. A detailed analysis of the provisions of the treaty is as follows:

Use of Terms

Article 1 describes the region of Southeast Asia and all the important terms mentioned in the treaty such as 'territory', 'nuclear weapons', 'station', 'radioactive material' and 'wastes and dumping'.

i) *Application of the Treaty*: Article 2 states that the treaty will be applicable to "territories continental shelves and Exclusive Economic Zones of state parties", but does not prejudice "rights of the exercise of these rights by any state under provisions of the law of the sea."

ii) *Basic Undertakings:* Article 3 describes the basic undertakings which says, not to develop, manufacture, acquire, possess, station, transport, test or use nuclear weapons and also not to allow any such action in its territory. It also prohibits dumping or disposing of radioactive material and waste.

iii) *Use of Nuclear Energy for Peaceful Purposes:* According to Article 4 nothing in this treaty shall prejudice the right of state parties to use nuclear energy for peaceful purposes. However, it directs the state parties to go through "rigorous nuclear safety assessment" recommended by IAEA before embarking on such a programme. It also calls for making available necessary information to state parties related to such programme, and to dispose of radioactive material and waste according to IAEA standards and also not to provide source or special fissionable material to any state except under conditions subject to NPT and IAEA.

iv) *IAEA Safeguards and Early Notification of a Nuclear Accident:* Article 5 invites for the member states to conclude an agreement with IAEA for the application of full scope safeguards and Article 6 desires them to accede to Convention on Early Notification of a Nuclear Accident.

v) *Foreign Ships and Aircraft:* As per Article 7 it was left to the state parties whether or not to allow foreign ships or aircrafts to visit their territory on being notified.

vi) *Establishment of a Commission and the Executive Committee*: According to the provisions of the Treaty, a Commission (as per the provisions of Article 8) and a subsidiary organ, the Executive Committee (Article 9) shall be established to oversee the implementation of the treaty and to ensure compliance with its provisions.

vii) *Control System*: According to Article 10, a control system shall be established for the purpose of verifying compliance with the obligations. The control system shall comprise IAEA safeguards as provided in Article 5, report and exchange of information (Article 11), request for clarification (Article 12), and request and procedures for a fact- finding mission (Article 13).

Other than these aforementioned provisions, Articles 14 to 22 deal with remedial measures, entry into force, reservations, relations with other international organisations, amendments, review, settlement of disputes and duration and withdrawal. The treaty also contains an Annex which deals with the procedure for a fact finding mission as provided for in Article 13.

Other than these provisions which are adherent towards state parties there is a separate chapter on protocols which has got seven articles. Some of these articles of the protocol are open for signatures by China, Russia, France, the United Kingdom and the United States. By signing the articles, signatories will assume various obligations which include to respect the treaty and not to contribute to any act that would constitute its violation, not to use or threaten to use nuclear weapons against any state party to the treaty and, in general, within the zone. The protocol is of permanent nature, but any party can withdraw from it, if it decides that extraordinary events related to the subject matter of the protocol have jeopardised its supreme interests. However, it may be mentioned that SEANWFZ has one specific feature which is not there in the case of South Pacific Nuclear-free Zone Treaty, and that is reaffirmation in the importance of NPT and its encouragement to establish nuclear weapon-free zones.

Assessment and Prospects of the Treaty

The United Nations Organisation supported the nuclear weapon-free zone in Southeast Asia as majority of General Assembly members supported the treaty. Former UN Secretary General, Boutros-Boutros Ghali, praised the resolve of Southeast Asian states to effect a treaty based denuclearisation (Activity Report of the Secretary-General of United Nations Boutros Boutros Ghali presented on 22 August 1996). The NPT Review and Extension Conference also praised the efforts made by the Southeast Asian states towards the establishment of SEANWFZ.[38]

However, all the five nuclear weapon states, till date, have refused to sign the protocol of the treaty and have raised several objections. They have complained about ASEAN's failure to consult them adequately prior to the signing of the treaty, a charge which ASEAN has rejected. The United States has all along been opposed to ZOPFAN and has indicated quite firmly to the ASEAN countries that it would not support the SEANWFZ proposal.[39] Since, the US has already signed the South Pacific Nuclear Free Zone Treaty and a similar treaty in Latin America it has made it clear that it does not object to the concept of nuclear weapon-free zone in Southeast Asia,[40] but has refrained from formally signing the protocol of the treaty.

The US position could be understood keeping in view the strategic balance of power and its military base interests which had existed in 1980s. However, in the present context, as the United States has already withdrawn from its bases in the Philippines and has no direct security concerns left in the region still it has opposed the treaty of SEANWFZ. The objections raised by the United States have been legalistic ones, such as the inclusion of continental shelves and Exclusive Economic Zones.[41] These objections have been raised on the ground of the consistency of the principles as embodied in the 1982 United Nations Convention on the Law of the Sea, mainly relating to high seas.

Another American concern involves the so-called negative security assurance, which would oblige Washington to pledge not to launch nuclear weapons from within the zone, even if it faced nuclear attack from another nuclear power that had signed the protocol. The US also finds the wordings of the provisions, guaranteeing the rights of ships and aircrafts carrying nuclear weapons to make local port calls and pass through the waters or airspace of the zone, very ambiguous.[42]

The fallout of the US not signing the protocol of the treaty would be that Britain and France would probably follow the US, as the US had signed the SPNFZ only when France had completed its tests in the South Pacific and was ready to sign the treaty.[43] However, on a visit to Singapore in early 1996, the French President Jacques Chirac stated that his country was 'predisposed to sign the treaty'.[44]

Russia has also failed to sign the SEANWFZ so far although it has signed several other such treaties like South Pacific Nuclear-free Zone. However, the fact remains that SEANWFZ does not do anything to restrict Russia's maritime possibilities and intentions in the region of Southeast Asia. The Russian Ambassador on a visit to Indonesia expressed Russia's support for the treaty 'in principle' but wanted some clarifications on how the treaty will be implemented.[45]

As far as China is concerned, it has been supportive of such initiatives right from the very beginning. In fact, as we have discussed in Part I, it has been the only nuclear weapon state to vote in favour of nuclear weapon-free zones in the United Nations. But in the case of Southeast Asia apart from raising objections about Exclusive Economic Zones and continental shelves,[46] China also objects to this treaty since it might affect China's existing claims on the Spratley islands.[47]

Beijing makes claims over all of the potentially resource-rich Spratley islands in the South China Sea and does not want the treaty to extend to continental shelves or Exclusive Economic Zones. In South China Sea, Brunei, Malaysia, Taiwan, the Philippines, Vietnam and China have overlapping claims to the Spratleys, one of the region's potential flash points. It has also been speculated in this regard that one of the motives behind ASEAN's desire to include Exclusive Economic Zones and continental shelves in the Zone is to restrict China from deploying or stationing nuclear weapons in the South China Sea.[48] Thus, in the case of China, signing of SEANWFZ appears to be difficult unless the issue of conflicting claims in the South China Sea is resolved.

The parties to the treaty, on the other hand, assert that the interests of the nuclear-weapon states have been adequately addressed by the treaty. They have confirmed that they do not want to violate the international law of sea. Moreover, the SEANWFZ is not expected to affect passage through the Strait of Malacca. Also the 'escape clause' which gives the right to an individual state whether or not to allow entry of nuclear powered ships and aircrafts into its territory, means the nuclear powers can have full access to port facilities and sea-lanes in the region even as the nuclear free zone is maintained.[49]

In view of the above the ASEAN members have appealed again and again to the nuclear weapon states to sign the protocol. Indonesian foreign minister Ali Atatas said in a meeting of ASEAN, "Since the treaty is not meant to undermine the security policies of the nuclear weapon states but rather to achieve security for all the countries in the region, we hope that the nuclear powers will endorse the treaty" and also "we have not lost hope that the nuclear weapon states will finally put a higher priority to their non-proliferation and disarmament commitments than their strategic postures and status as nuclear powers."[50]

Since support of the nuclear weapon states are important, ASEAN agreed to a dialogue with the United States to accommodate its reservations. The United States has proposed either to delete all the references about continental shelves or to issue an 'interpretative statement' that there provisions would be applied only to the state parties.[51]

SEANWFZ has remained high on ASEAN's agenda and has been taken up in all the ASEAN Regional Forum (ARF) meetings. ASEAN wants that the ARF endorses the treaty of SEANWFZ. A question, however, may arise in this regard as to how much nuclear threat the region presently faces and whether the treaty has been able to fully address the threat. Southeast Asia which is a neighbouring region of India and Pakistan, the nuclear threat does exist but the treaty provisions do not take this into account at all or seek any obligations from India or Pakistan. Another issue of concern in this regard is China's expanding nuclear arsenals and North Korea's nuclear programme. But this issue has also not been touched upon.

Another question which could be raised in this regard is that whether removal of the nuclear threat alone will make any significant contribution to the security of the region. Unlike the South Pacific, Southeast Asia does not face any direct nuclear threat such as nuclear testing or dumping of nuclear wastes. In this case a NWFZ in Southeast Asia can only be of limited significance apart from the political, psychological and symbolic benefits that can be derived from engaging in a process that has the potential in the long term to influence the strategic environment in Southeast.[52]

Thus the security benefits to be gained by the treaty of NWFZ in Southeast Asia are going to be marginal and it is unlikely to make any significant impact on the resolution of security issues currently confronting the countries of the region.

Benefits can be seen, however in the political and diplomatic realm where the agreement can be used to regulate the conduct of nuclear weapon states which goes with the spirit of ZOPFAN and in turn may have the potential to modify the strategic environment of Southeast Asia.[53]

REFERENCES

1. See *Text of Kuala Lumper Declaration*, appendix I.
2. For details see Roeslan Abdulgani, *The Bandung in 1955*, (Singapore: Gunung Agung, 1981).
3. Text of the 1955 Bandung Conference Published in *Keesings Records of World Events* (UK.: Longman 1955), pp.14181-14184.
4. Habib, Hasnan, "ASEAN in Search of Peace and Stability in Southeast Asia", United Nations Disarmament Work-shop for Asia-Pacific (New York: UN, 1991, p. 231.
5. See *Neutralisation of Southeast Asia: Problems and Prospects* (Washington: US Government Printing Office, 1976).
6. Alagappa, Muthia, "Security in Southeast Asia: Beyond ZOPFAN", *Topical Paper* 6 (New York: United Nations, 1991), p.112.
7. Ibid., p.119.
8. Ibid., p.111.
9. Ibid., p.112.
10. Negi, R., *ASEAN: 20 Years* (New Delhi: Lancers Books, 1989), p.36.
11. Miller, T.B., "The Five Power Defence Arrangement and Malaysian Security", *Pacific Community*, Vol.3, No.2, January 1972, p.343.

12. M. van der Kroef Justus, "ASEAN's Security Needs and Policies", *Pacific Affairs*, Vol.47, No.2, Summer 1974, p.159.
13. Cited in Seodjati Djiwandono, "Cooperative Security in the Asia Pacific Region: An ASEAN Perspective", *The Indonesian Quarterly*, Vol. 22, No.3, 1994, pp.6-67.
14. Alagappa, Muthia, "Regional Arrangement and International Security in Southeast Asia: Going Beyond ZOPFAN *Contemporary Southeast Asia*, Vol. 12, No.4, March 1991, p. 281.
15. Horn, Robert C., "The Soviet Perspective", *Southeast Asia under New Balance of Power* (New York: Praeger Publishers, 1975), p.33.
16. *Times of India*, 8 August 1975.
17. Chee, Chau Heng, "Southeast Asia 1976: The Handling of Contradiction", *South East Asian Affairs*, (Singapore, ISAS, 1977), p.20.
18. Djiwandono, Seodjati, "Neutralisation: A New Hope for Southeast Asia?", *The Indonesian Quarterly*, Vol.I, No.2, January 1973, p.67.
19. Leifer, Michael, *ASEAN and the Security of Southeast Asia* (London: Routledge, 1989), p.7.
20. Denoon David B.H. and Colbert Evelyn, "Challenges for the Association of Southeast Asian Nations (ASEAN)", *Pacific Affairs*, Vol.71, No.4, Winter 1998-99, p.507.
21. See the Text of Bali Summit published in Negi, n.10.
22. See Negi, n.10, pp.128-34.
23. Roeslan Abdulgani, Nationalism, Regionalism and Security: Problems in Southeast Asia (New Delhi: Banyan Pub), p. 50.
24. See Article 1. *Text of the Treaty of Amity and co-operation*, appendix III.
25. Acharya, Amitav, "ASEAN-UN Co-operation in Peace and Preventive Diplomacy: Its Contribution to Regional Security", *The Indonesian Quarterly*, Vol.22, No.3, 1994, p.221.
26. *Los Angles Times*, 15 April 1974.
27. *Far Eastern Economic Review*, Asia 1977 Yearbook (Hongkong, 1977), p.230.
28. Acharya, Amitav, *The Quest for Identity: International Relations of Southeast Asia* (Singapore: Oxford University Press, 2000), p.109.
29. Das, Parimal Kumar , "Security Environment in Southeast Asia" in P.K. Das, (ed.), *The Troubled Region* (New Delhi: Sage Pub., 1987), pp.37-38.

30. The London suppliers group and Zagger Committee, "Southeast Asian Nuclear-Weapon-Free Zone: Ideas, Problems, Prospects and Proposals", *UN Regional Disarmament Workshop for Asia Pacific* (New York: United Nations, 1991), p.201.
31. Snitwongse, Kusuma, "Southeast Asian Security Issues and Confidence Building", *UN Regional Workshop for Asia Pacific* (New York: UN, 1991), p.144.
32. Alagappa, Muthia, "A Nuclear Weapons Free Zone in Southeast Asia: Problems and Prospects",*Asutralian Outlook*, Vol.41, No.3, December 1987, p.174.
33. Muetzenich, Rolf, "Nuclear Weapon Free Zone in Southeast Asia", *Aussen Politik*, Vol. 48, No. 4, 1997, p. 393.
34. *Beijing Review*, 1 June 1996, pp.17-20.
35. *Hindustan Times*, 19 December 1995.
36. Jr. Abad, M.C., "Towards an ASEAN Common Foreign and Security Policy", *Indonesian Quarterly*, Vol.XXVII, No.1, First Quarter, 1999, p.5.
37. See Amitav Acharya, n. 25.
38. *SIPRI Yearbook 1996*: Armaments, Disarmaments and International Security, 1996, p.575.
39. Secretary of State George Shultz wrote to Indonesian Foreign Minister Mochtar Kusumamatmadja, *International Herald Tribune*, 25 June 1986, p.2.
40. *International Herald Tribune*, 23 July 1996.
41. Ibid.
42. Pande, Savita, "Prospects of Nuclear Weapon Free Zones: Southeast and Central Asia", *Strategic Analysis*, Vol.22, No.6, September 1998, p.863.
43. France likely to Sign "S-E Asia N-Pact Soon', *Strait Times*, 3 March 1993, p.18.
44. FBIS-SOV-97-208, 27 July 1997.
45. http:#www.cdis.org
46. *The Pioneer*, 16 December 1995.
47. *International Herald Tribune*, 23 July 1996.
48. *Times of India*, editorial, 15 December 1995.
49. *Deccan Herald Tribune*, 22 July 1996.

50. Acharya, Amitav , No.1, p. 446-47.
51. Alagappa, Muthia, “A Nuclear Weapons Free Zone in Southeast Asia: Problems and Prospects”, *Australian Outlook*, Vol.41, No.3, December 1987, p.175.
52. Ibid., p.179.
53. Ganeshan, N., “ASEAN's Relations with Major External Powers”, *Contemporary Southeast Asia*, Vol.22, No.2, August 2000, p.272.

7

Regional Security in Southeast Asia and the Relevance of ZOPFAN

THE ZOPFAN IDEA

During the first special meeting of the ASEAN Ministers for Foreign Affairs held in November 1971, adopted the Kuala Lumpur Declaration on the establishment of Zone of Peace, Freedom and Neutrality (ZOPFAN) in Southeast Asia. It was a declaration by the ASEAN member countries to take charge of their own destiny and reject the assumption that the fate of their region was to be determined by outside powers. It also reflected ASEAN's strong belief that it was only through close cooperation, relying primarily on indigenous strength, that peace security and stability in their own region may become the primary responsibility of the Southeast Asian nations themselves.[1]

What does ASEAN understand by a Zone of Peace, Freedom and Neutrality? As Dr.Roeslan Abdulgani explained it "a Zone of Peace Freedom and Neutrality exits where the national identity, independence and integrity of a state within such a zone can be preserved and maintained, so that it may achieve its national development and well-being and also promote regional cooperation and solidarity in accordance with the ideals and aspirations of its people and the purposes and principles of the UN charter without any form or manner of interference by outside powers".[2]

Dr. Abdulgani further explained the definition of peace. He has written, "peace is a condition characterised by harmonious and orderly relations between and among states: no reference is hereby made to the internal state of affairs in each of the state in the zone. A situation of conflict. Ideological, political economic, armed or other, either among the states of the zone themselves or between one or more of the states of the zone and outside powers, or between outside power affecting the region is not a condition of peace".[3]

Describing the meaning of 'Freedom' Dr. Abdulgani has written, "Freedom means the freedom of state from control domination or interference by other states in the conduct of their internal and external affairs. This means the right of states of the zone to solve their domestic problems in terms of their own conditions and aspirations to assume primary responsibility for region, and to arrange their regional and international relations on the basis of sovereign equality and mutual benefit".[4]

Lastly, Neutrality means the "maintenance of state of impartiality in any war between other states as understood in international law and in the light of the UN charter. However, in the context of the Kuala Lumpur Declaration, it means that the states of the zone undertake to maintain their impartiality and reform from involvement, direct or indirect in ideological, political, armed or other forms of conflict, particularly between outside powers and are not allowed to interfere in the domestic or regional affairs of the states of the zone".[5]

ZOPFAN AND CAMBODIAN INVASION

As discussed earlier the fears of Vietnamese potential for a military intervention in the region proved to be true. On 28 December 1978, only months after the signing of a twenty years 'friendship treaty' with the Soviet Union, Vietnam invaded Cambodia and deposed the governing Khmer Rouge, Hanoi's intervention brought a speedy reaction from China. It launched a counter attack from northern borders with an aim of 'teaching Hanoi a lesson'. In turn Vietnam became more dependent on the Soviet bloc, thus the battle lines were clearly drawn with USSR and Vietnam on one side and the US and China on the other. The US interests in Southeast Asia which had declined since 1975 appeared to have been revived. Writing about US interest in Southeast Asia in post 1975 period, one of the scholars has written that "the American interests since then were mainly guided by ASEAN and China. In effect it said to ASEAN and China—you lead and we follow, any policy mutually accepted to you as acceptable to us."[6]

But the Vietnamese Occupation of Cambodia sharply reawakened American interest in the region. This marked yet another return to conventional power politics into the region and gave a major blow to ZOPFAN idea. In fact the very situation arising from Indo-China conflict which had actually prompted ASEAN countries to adopt ZOPFAN, in effect, became a major impediment in its realisation. In the words of J. Saravanamuttu, "It is undeniable that the Indo-China developments have dealt a mortal—some would suggest fatal blow to the ZOPFAN idea."[7] To quote Amitav Acharya, "...the third Indo-china war aggravated the security dilemma in Southeast Asia and further undermined ASEAN's professed security framework (especially ZOPFAN)".[8]

The invasion also came as a shock to ASEAN. Thailand faced the greatest security threat in the wake of being the frontline state. "The occupation of Kampuchea has posed a historically unprecedented threat to the security of a state which was never subject to direct colonial domination".[9] However, there was little that it could have done to prevent the Vietnamese invasion of Cambodia. There were also differences in ASEAN over the course of action to be adopted against Vietnam. Although ASEAN members maintained a total solidarity on face regarding their attitude towards Vietnam. There existed serious difference among them. Thailand considered Vietnam as a historical rival and wanted a neutral Cambodia as a buffer state between itself and Vietnam. Therefore, it improved its relations with Vietnam's enemy China. Singapore shared Thailand's viewpoint. Indonesia and Malaysia on the other hand regarded China as a greater enemy for the security of the region.[10]

To quote V. Suryanarayan in this regard, "From the beginning the ASEAN unity was more apparent than real as the six countries differ in their own assessment of Vietnam and China, depending on their national interests and their appraisal of the geopolitics of the region in the light of Sino-Vietnamese and Sino-Soviet dispute."[11]

The situation had not only led to ASEAN's involvement in the Sino-Soviet rivalry but also opening Southeast Asia to a wider expression. Moreover, Vietnamese aggression was against the sanctity of national sovereignty, enthroned in the Treaty of Amity and Co-operation. However, one single effort towards spirit of ZOPFAN came in the form of 'Kuantan Declaration'. The Malaysian Prime Minister Datuk Hussein met with Indonesia's President Suharto in March 1980 in the coastal town of Kuantan in Peninsular Malaysia and their discussion led to the enunciation of 'Kuantan principle'. The crux of the declaration was that ASEAN wished to dissociate itself from China's announced policy of 'bleeding Vietnam white', while seeking to convince Moscow that it was in the Soviet interest to have better relations with ASEAN.[12]

But the Kuantan message was allowed to lapse due to Constant hostility shown by Thailand towards it. Anyhow it did show that there existed some internal dissentions in ASEAN over the Cambodian crisis. In this regard, it may be stated that ZOPFAN was seriously undermined by ASEAN's stand. Its encouragement of Chinese and American involvement in Southeast Asia to balance Vietnamese and Soviet activity was in contradiction to its own expressed principle of neutrality, but it did assert to uphold the principles of treaty of Amity and Co-operation,[13] and ASEAN continued to take a unified stand on the issue in almost all the international fora. ASEAN states successfully used their membership of the United Nations, the Organisation of Islamic Conference and the Commonwealth as well as dialogue relations with the US, EC, Japan, Australia, Canada and New Zealand to mobilise international support for their cause, and kept international diplomatic and economic pressure on Vietnam. And finally after much wrangling and haggling not unknown to the world, Indo-China conflict reached a settlement. Vietnam's forces were withdrawn from Cambodia; there was also a normalisation of US-Vietnam relations and ultimately cold war ended with the demise ofUSSR.

Thus, ASEAN's role in bringing about peace in the region has been valuable however as has been stated by Muthia Alagappa the ASEAN experience shows that regional organisations can play only a limited role in preventing conflicts involving non-member regional states as well as extra-regional states.[14] This is evident from ASEAN's limitation in excluding great power rivalry and settling related conflicts in Southeast Asia. But had ASEAN been successful in doing so it would have certainly have facilitated the realisation ofZOPFAN.

REGIONAL SECURITY IN SOUTHEAST ASIA AND THE RELEVANCE OF ZOPFAN

With the end of Cold War, and disintegration of Soviet Union several changes occurred on the international scene. The world saw political settlement of several of the long-standing conflicts abbeted by the cold war. The war in Indo-China is one of such conflicts which finally ended with Vietnamese withdrawal from Cambodia. It was no doubt the major hurdle in the realisation of ZOPFAN and had made the region a 'cockpit' of great power rivalry and had also polarised the regional countries. However, the settlement of the Cambodian conflict did not put a complete end to the problems in Southeast Asia, other threats to regional stability, such as territorial disputes in the South China Sea came up. The South China Sea is, as we are aware, one of the most strategic waterways in the world and also has enormous living and non-living resources.

There are two groups of disputes in the South China Sea, the first group of disputes exists between China and Vietnam over parcels; and the second exists among six countries over Spratleys, which are China, Taiwan, the Philippines, Vietnam, Malaysia, Brunei and Indonesia. It may be noted in this regard that the dispute gained prominence in late 1980s. In 1987 China's navy constructed observation stations in the Spratley islands and the Chinese military forces clashed with the Vietnamese forces in the area. In 1992, the Chinese People's National Congress declared that the Spratleys were a vital part of China and subsequently occupied five islands and also installed oil drilling platform in the disputed areas. In 1995, China seized the Mischief Reef in the Spratleys and Chinese maps now include most of the South China Sea as part of China's sovereign territories.

The dispute in the South China Sea is one of the greatest challenges for regional security and ASEAN is, faced with the undaunted task of finding a harmonious solution to this problem. In view of this, ASEAN has organised a series of annual informal workshops over this issue. There is also a concern about a possible armed conflict breaking in the region. As a matter of fact, ASEAN has always viewed its northern neighbour with much apprehension. After the cold war and withdrawal of the US forces from its bases in Southeast Asia there arose concerns about a long-term shift in balance of power in China's favour. Thus, China has become a crucial factor in the security architecture of Asia-Pacific and ASEAN had maintained a constant focus on China for many reasons. While China is an important partner for trade and investment in the region, its expansionist policies have given rise to constant fear.[15]

However, ASEAN so far has not displayed any anti-China position and instead it has tried to engage China consistent with its policy of "constructive engagement". For ASEAN, a good relationship with China is vital but at the same time ASEAN has tried to create a balance of power in the post cold war era, with a system of cooperative security arrangement which would involve the low-key presence of four or five great powers in the region—the US, China, Japan, Russia and India —to prevent a hegemony emerging.[16] One such effort in this regard has been the creation of Asian Regional Forum (ARF) in 1994. Presently, the forum has 21 members which are important players in the Asia Pacific region. The ARF had been instilled with three main objectives—confidence building measures (CBMs), preventive diplomacy and the question of non-proliferation and arms control. These objectives are, more or less, similar to those of ZOPFAN. In the case of South China Sea disputes also, the confidence building measures and preventive diplomacy have been taken into account. At the Seventh ARF meeting the Chairman stated that "With regard to the South China Sea, the Ministers encouraged the exercise of self restraint by all countries concerned and promotion of confidence building measures."[17]

In this regard, a range of confidence building and preventive diplomacy measures has been taken up by ASEAN. The confidence building measures could be manifested in Treaty for SEANWFZ. One of the methods by which the concept of preventive diplomacy has been advanced is through the dialogue partners of ASEAN have who endorsed the Treaty of Amity and Cooperation. A formal treaty based code of conduct, such as those embodied in the Treaty of Amity and Cooperation, has been commissioned by ASEAN in the South China Sea to defuse tensions and turn the focus on joint development efforts.[18]

The mechanism provided by the Treaty of Amity and Cooperation to resolve the issues relating to regional countries own outstanding territorial claims forms the basis of the code of conduct. In 1992 the United Nations Organisation's resolution on Treaty of Amity and Cooperation was adopted by consensus in 47th General Assembly.[19]

The ARF in its first session held in Bangkok (25 July 1994) agreed to accept the objectives and principles of the Treaty of Amity and Cooperation a code of conduct governing relations between states and a unique diplomatic instrument for regional confidence building, preventive diplomacy, and political and security cooperation (paragraph 6).

At the 6th ARF in Singapore (25 July 1999) members recognised the Treaty of Amity and Co-operation "a key regional instrument for strengthening security in the region". This will promote the development of a "region-wide code of conduct". Moreover, the acknowledgement of non-Southeast Asian countries of the SEANWFZT is expected to contribute to the comprehensive denuclearisation of the region.[20]

As far as non-proliferation and arms control is concerned, ASEAN aims at promoting concepts of ZOPFAN, SEANWFZ and TAC. To quote Jusuf Wanandi, ASEAN wants a central role in ARF because ASEAN has relevant ideas of cooperative security such as Zone of Peace, Freedom and Neutrality (ZOPFAN) and the Southeast Asian Nuclear Weapon Free Zone (SEANWFZ) to apply and is itself a successful multilateral institution.[21]

ASEAN as ASEAN-10 is better equipped to approach the problem confronting the region in a more comprehensive manner. It is expected that the problems like South China Sea and other bilateral disputes such as over Sabah between Malaysia and the Philippines; the Thai-Malaysian border problems and Thai-Cambodia dispute over oil reserves in the Gulf of Thailand would be dealt in a better manner keeping alive the 'ASEAN WAY' and following the spirit as proposed in the ZOPFAN.

It may, however, be said in this regard that ZOPFAN, as it was proposed in 1971, would not be relevant for Southeast Asia as it is. In this age of cooperative security, there is greater acceptance that security in Southeast Asia can be best enhanced through the active cooperation and engagement of relevant outside powers. It is more about engaging the external powers in a constructive manner rather than proposing for their neutrality.

But the Treaty of Amity and Co-operation and SEANWFZ as instruments of ZOPFAN are still relevant to the region and the current moves to enhance regional security co-operation in Southeast Asia recognise the need for strengthening of existing mechanisms like TAC and SEANWFNZ as well as development of additional measures.

REFERENCES

1. Hasnan Habib, "ASEAN in the Search for Peace and Stability in Southeast Asia", *Regional Disarmament Workshop for Asia Pacific* (New York: United Nations, 1991), p.231.
2. Roeslan Adbulgani, *Nationalism, Regionalism and Security: Promblem in Southeast Asia* (New Delhi: Banyan Pub.), p.47.
3. Ibid., p.47.
4. Ibid., p.48.
5. Ibid., p.48.
6. Douglas Pike, "Southeast Asia and the Super Powers: The Dust Settles", *Current History*, Vol.82, No.483, April 1983, p.146.
7. J. Saravanamuttu, "ASEAN'S security for 1980s: The Case for a Revitalised ZOPFAN", *Contemporary Southeast Asia*, Vol. 6, No. 6, September 1984, p.189.
8. Amitav Acharya, *The Quest for Identity: International Relations of Southeast Asia* (Singapore: Oxford University Press, 2000), p. 117.
9. Michael Leifer, *ASEAN and the Security of Southeast Asia* (London: Routledge, 1989), p.90.
10. See D.R. Sardesai, Southeast Asia: Past and Present (New Delhi: Vikas Pub. 1981).
11. V. Suryanarayan, "Settlement in Sight", *World Focus*, Vol.9, No.6, June 1988.
12. Ronald G. Palmer and Thomas J. Rockford, *Building ASEAN : 20 Years of Southeast Asian Co-operation,* (Washington; Centre for strategic and International Studies, 1987), p.70.
13. Shaun Narine, "ASEAN and the Management of Regional Security", *Pacific Affairs*, Vol.71, No.2, Summer 1998, p.207.
14. Muthia Alagappa, "Regionalism and Quest for Security: ASEAN and the Cambodian Conflict", *Australian Journal of International Affairs*, Vol.47, No.2, October 1993, pp.203-204.

15. N. Ganeshan, "ASEAN's Relations with Major External Powers", *Contemporary Southeast Asia*", Vol. 22, No.2, August 2000, p. 272.
16. Jusuf Wanandi, "ASEAN's China Strategy: Towards Deeper Engagement", *Survival*, Vol.38, No.3, Autumn 1996, p.120.
17. *The Hindu*, 30 July 2000.
18. *The Hindu*, 28 July 1999.
19. Taken from Abdul Samad Paridah and Muhammad Mokhtar, "ASEAN'S Role and Development as a Security Community", *Indinesian Quarterly*, Vol 23, No. 1, 1995, p. 72.
20. Taken from C.P.F. Luhulima, "ASEAN's Relevance: Has it Become Questionable", *Indonesian Quarterly*, Vol.XXVII, No.4, 1999, p.297.
21. Jusuf Wanandi, *ASEAN's China Strategy*, No.15, p.121.

PART III

Nuclear-free Zone in the South Pacific

8

The South Pacific and Its Security Concerns

On 6 August 1985, the South Pacific became the second inhabited region, after Latin America, to be declared as a nuclear free zone. This was achieved by a treaty signed during a meeting of the South Pacific Forum,[1] in Rarotonga in the Cook Islands. Eight of the thirteen members, namely, Australia, New Zealand, Western Samoa, Tuvalu, Nive, Fiji, the Cook Islands and Kiribati, signed the treaty on the same day, while the other five members—Tongo, Papua New Guinea, Vanuatu, Nauru and Solomon Islands—endorsed the treaty and were soon expected to sign it. The pact was signed on 6 August 1985, which was the 48th anniversary of Hiroshima bombing. Paul F. Power, commenting on the pact, stated that for the independent South Pacific nations, which were already free of nuclear weapons before the treaty emerged, "the Rarotonga Pact is a step onto the global anti-nuclear stage".[2] The treaty also reflected a strong and long-standing anti-nuclear sentiment of the South Pacific nations and their opposition to nuclear testings by the nuclear weapon states such as the United States, Britain and France as well as dumping of wastes in their sea.

The Treaty of Rarotonga, as the South Pacific nuclear-free-zone treaty is generally known, developed from within the region, as a desire of the South Pacific countries to free their region from the nuclear madness. It was a product of South Pacific foreign policies and politics dating back to the 1960s.[3]

The growth of anti-nuclear sentiments led to a movement which was committed to the nuclear-free Pacific. In fact, the South Pacific Island-states have long been concerned about nuclear activities taking place in their region. Since the dawn of the nuclear age itself, the South Pacific faced a direct threat from nuclear activities such as nuclear testing, dumping of nuclear wastes, presence of nuclear installations and transit of nuclear powered ships, turning it into one of the most nuclearised regions of the world.

Nuclear weapon states, such as the United States, the United Kingdom and France which are also the former colonial masters and still have several bases in the area, took advantage of the remoteness, smallness and tractability of widely dispersed population to turn the pacific paradise into a nuclear hell.

However, before proceeding further, it would be appropriate to make some observations about the general atmosphere and internal situation of this relatively lesser known region. The south Pacific region, as it is being discussed here includes the area covering all member countries of South Pacific Forum, Australia, New Zealand, Papua New Guinea (PNG), Fiji, Vanuatu, Kiribati, Western Samoa, Tuvalu, Tonga, Nive, the Cook Islands and Solomon Islands, (Forum Island Countries (FICs) here onwards).

The South Pacific is a vast region and it stretches 17,000 km longitudinally and 7,000 km latitudinallly. Less than 3 per cent of this area is land which is fragmented into thousands of islands. However, not all of these islands are inhabited. Apart from two 'metropolitan' countries of Australia and New Zealand there are nine independent sovereign states in this region, and these are Western Samoa, Nauru, Fiji, Tonga, Papua New Guinea (PNG), Solomon Islands, Tuvalu, Kiribati (pronounced as Kiribass) and Vanuatu.

Apart from these countries, there are four self governing states, namely Nive and Cook Islands in Free Association with New Zealand, Federated States of Micronesia, and the Republic of Marshall Islands in Free Association with the United States of America. Apart from Fiji and PNG, all the countries of the South Pacific fall under the category of 'Microstate'. However, they vary enormously in regard to both size and population.

Geologically also, they vary from high volcanic islands as in Rarotonga to low coral atolls varying in size to larger continental islands such as PNG, Fiji and New Caledonia which are rich in minerals. Except for the larger continental islands most of the island states of the South Pacific have fragile eco-systems. All of them are vulnerable to natural disasters such as hurricanes, volcanic eruptions and earthquakes. The geographical diversity of this region is also matched by the diversity of people and their culture who can be divided into three broad categories: Melanesian, Polynesian and Micronesian.[4]

There is a considerable diversity among these three cultural groupings but one common factor which is found homogeneously throughout the region is an atmosphere of relative peace and harmony. Unlike most of the other regions of the world, since 1945 the use or threat to use of armed forces in shaping intra-regional relations had been absent in this region. As a consequence, the defence capacity in most of the Forum Island Countries had been very low to non existent. As a matter of fact, apart from Australia and New Zealand, only PNG and Fiji maintain substantial defence capacity.[5]

As far as external military threats are concerted, except for the period of World War II when these islands were used by Japan as well as the adversaries for different purposes, they have been largely absent. Other than their relations with trans-Pacific lanes of communications, these island states were perceived as having little strategic importance.[6] As noted scholar on the area Ken Ross describes, these island states, "constitute one of the world's most democratic and stable regions and is also of marginal global strategic importance".[7]

But as these island countries became independent a renewed interest of great powers could be seen. This interest was further increased by the declaration of the 'Law of the Sea Convention in 1982 which permitted island states to claim rights over their extensive 'Exclusive Economic Zones'.[8]

Nevertheless, these islands continued to be low on the priority of western powers, especially the United States which was the major external power in this region. It is the former colonial power which maintains its presence through regional security pact ANZUS with Australia and New Zealand. Another colonial power which still has a significant presence in the region is France. It has New Caledonia as its colony and has conducted maximum number of nuclear tests in the region.

A new entrant to the list of very active external powers is Japan. Although Japan's attempt to conquer the Pacific militarily during the World War II was quite disturbing but now this is a matter of little concern to many and Japan has established its embassies or consulates in several of the island states while some of the Forum Island Countries have trade offices in Tokyo. Japan has maintained trade relations with these countries, however, balance of trade has overwhelmingly been in Japan's favour. The main imports of Japan being PNG's copper, New Caledonia's nickel, Solomon Island's timber, and fish caught under licence in most islands exclusive economic zones. As regards Japan's exports, Japanese vehicles and electronic equipment etc., have virtually taken over the island markets. Japanese aid is substantial and Japanese experts, volunteers, businessmen, tourists and researchers are now evident throughout the region.[9] Now the countries such as China, Taiwan and some of the ASEAN countries are also involved in the region but their interest remains more or less commercial.[10]

Thus, it can be seen that the interest of external powers in South Pacific was not exactly linked to any kind of military involvement. Also, there was no direct threat to the region in the conventional sense, as there was no military threat or any fear of being attacked by either any regional country or by any outside power. But this is not to say that 'trouble spots' have not been there at all. The security concerns have been very much there but have a non-conventional dimension.

In the virtual absence of external and intra-regional military threats, primarily non-military security issues have pre-occupied the Forum Island Countries such as economic uncertainties, environmental degradation, law and order problems, resource management problems, internal unrest and secessionist movements. In this regard, it is important forms to discuss these problems as they are closely linked to the regional security issues.

THE ETHNIC CONFLICT IN FIJI

Fiji is a small island country which was a British colony till 1970. The Britishers in want of cheap labour imported large number of Indians under indentured labour system to work on their sugar plantations. The two communities, ethnic Fijians and Indians were kept apart from each other by the Britishers. At the time of Fiji's independence Indians formed around 50 per cent of the population. The problem in Fiji stems from the fact that each of the two communities has difficulty in accepting the other. The Indians who are economically strong also want equal share in the politics of the country while the ethnic Fijians, who are economically and educationally still quite backward, are determined to rule the country in their own way.

As a result of this power struggle there were coups in 1988 and again in 2000 against Indian dominated governments of Dr. Timoci Bavadra and Mr. Mahendra Chaudhary respectively. Both the coups affected the general stability of the region. As regards the reaction of the Forum Island Countries, there was a clear division among them as the island nations adopted a more liberal attitude towards the incident than that of Australia and New Zealand who wanted to punish the military regimes which took control after toppling democratically elected governments.[11]

SECESSIONIST MOVEMENTS IN NEW CALIDONIA

New Calendonia is a French colony in the South Pacific. French policies kept the indigenous people known as Kanaks, out of the mainstream developmental process. They have remained educationally, socially and economically backward. At several occasions the Kanaks have revolted for facilities and finally independence, to which the French have not agreed to concede.

Several accords like Matignon and Naumea have been signed between the French and the indigenous Kanak people which gave Kanaks a larger share in the government, more financial benefits and assurance for complete independence in the future. As far as Forum Island Countries are concerned there has been an overwhelming support for Kanaks among them and the issue of New Caledonia has been one of the major causes of anti-French feelings in the region.[12]

THE PROBLEMS IN PAPUA NEW GUINEA (PNG)

There are two major problems in Papua New Guinea. The first problem is related to the western part of New Guinea or Iriyan Jaya which is under Indonesian control. A separatist movement is going on in Irian Jaya against Indonesia for quite some time. As the rebels cross the border and take shelter in PNG and very often Indonesian operations against these rebels also violate the sanctity of demarcated borders, there has been a constant border conflict between these two countries. However, the PNG government has tried to resolve this problem with Indonesia through dialogue and has managed to dilute it to some extent.

The other problem which has affected the otherwise benign security environment in the South Pacific is the Bougainville issue. Bougainville is a province of PNG. It is rich in copper which leads to extensive mining activities. Bougainvilleans are ethnically different from rest of the PNG people. They have resisted excessive mining activities in their province as these activities have proved disastrous for their environment and the benefits emanated from the mining have not exactly reached them. In view of this problem, a separatist movement is going on in the province. However, the PNG government has tried to suppress this movement. As regards other regional countries, Australia's position is the most precarious one.

It may be noted that geo-strategically, PNG is very important for Australia as presence of the Indonesian troops in the western half of PNG makes the situation vulnerable for Australia. Australia, on its part, has maintained its heavy presence in PNG by providing aid, assistance and military support. However on the one hand Australia is apprehensive of Indonesian motives and on the other, it has been criticised by the Forum Island Countries for its indiscriminate assistance as the PNG government has used a large share of this assistance against Bougainville rebels. This issue has been taken up in the South Pacific Forum at several occasions and has remained a burning issue in the region.[13]

ENVIRONMENTAL ISSUES

The issue of environment in the South Pacific has been viewed as one of the major aspects related to regional security. Environmental threats faced by these island states actually pose threat to their very existence. In most of the cases the damaging environmental problems such as global warming, hazardous waste dumping, resource depletion and driftnet fishing are caused by the countries outside the region. However, it was only in 1980s that these island countries began giving importance to environmental issues and accorded them significant priority in their national policy. In 1986, the convention and the protocols of Protection of Natural Resources and Environment of the South Pacific (SPREP) were signed. The preamble to the SPREP recognises the need for parties to cooperate among themselves and with competent organisations to ensure sustainable resource management.

Another environmental threat faced by the South Pacific Island countries was driftnet fishing. The system became common in 1980s and led to indiscriminate and excessive fishing. The matter was taken up very seriously by the island countries. Through regional organisations like South Pacific Commission (SPC) and South Pacific Forum (SPF) they took the issue to several international fora. Finally, the issue was taken to the United Nations Organisation and a resolution demanding it's ban was put forward but a compromise was reached with Distant Water Fishing Nations (DWFNs) which proposed a moratorium on large scale driftnet fishing and immediate reduction of such a method which lead to excessive fishing. Besides driftnet fishing, other environmental issues such as global warming and hazardous waste dumping were also taken up by small island countries from time to time.[14]

THE FISHERIES MANAGEMENT AND RELATED ISSUES

Tuna is the major commercial fisheries resource of the South Pacific. Tuna catches from the South Pacific form the major portion of world's total Tuna catch. However, most of the high seas catches are taken not by the Island countries to where exclusive economic zones these fishes belong, but by Distant Water Fishing Nations (DWFNs) which are mainly, United States, Japan, Taiwan, Korea, the Philippines etc. In fact, the Island countries lease out their waters to DWFNs in return of a fee which is usually determined on the basis of expected value of the catch per vessel. But over fishing and indiscriminate fishing is a common practice which goes against the interest of these small island countries. Another problem area was the United States position on Tuna. The United States refused to recognise coastal states' jurisdiction over highly migratory species of fishes such as Tuna unless coastal states cooperate in its management. The illegal catching of Tuna by the Distant Water Fishing Nations and particularly by the United States became region's most important environmental issue. In June 1984,

Solomon Islands confiscated American fishing boat in its waters and this event brought the fishery differences to climax. Consequently, an embargo was imposed by the United States on Solomon Islands which greatly affected its economy. This development enraged the island countries and finally, the United States had to lift the embargo. Eventually, a regional fisheries agreement was signed with the United States under the auspices of the Forum Fisheries Agency. It may be mentioned that the Forum Fisheries Agency is a regional organisation in the South Pacific which looks into fisheries related matters. This Agency has been a very successful organisation and has taken fisheries management in a big way.[15]

ECONOMIC VULNERABILITIES

Economic vulnerability is one of the immediate threats for most of the Forum Island Countries. Some of the general features of their economy include small markets, distance from major markets of the world, concentration of exports around a narrow range of basically primary products, unfavourable balance of trade, heavy dependence on aid etc. The major resources earners in most of the countries are remittances, aid and license fees for leasing out waters for fishing etc. To add to all these natural calamities are a source of constant concern. There are also the problems related to over-population and unemployment. Since all these countries depend heavily on aid and assistance they are sometimes not able to take decisions independently.[16]

Thus, from the above discussion it can be gathered that threat environment or security concerns of the Forum Island Countries are not limited to conventional military threat. It includes anything that might force them to compromise with their political and economic sovereignty and their control over regional development. In this regard, the nuclear issue which remained the major concern for the region till late 1990s, should also be seen in light of the threat as perceived by the Countries of the South Pacific. Unlike Southeast Asia, the nuclear threat, for the South Pacific was not exactly, of being attacked by nuclear weapon instead, the issue was of being used for nuclear testing or dumping of nuclear wastes and other kinds of nuclear activities by the great powers. In fact, the threat was posed to their environment, their sovereignty and their very existence. It is in this light that we would be discussing the nuclear threats in the region.

REFERENCES

1. South Pacific Forum is an organisation of all the independent states of South Pacific. Founded in 1971, its members are Australia, the Cook Islands, Fiji, Kiribati, Nauru, New Zealand, Nive, Papua New Guinea, the Solomon Islands, Togo, Tuvalu, Vanuatu and Western Samoa.
2. Paul F. Power, "The South Pacific Nuclear Free Zone", *Pacific Affairs*, Vol.59, No.3, Fall 1986, p.455.
3. Gregory E. Fry, "Regionalism and International Politics of the South Pacific",*Pacific Affairs*, vol.54, No.3, Fall 1981, pp.455-84.
4. Jim Sandy, "South Pacific Culture and Politics: Notes on Current Issues", *Working Paper No.174* (Strategic and Defence Studies Centre, Australian National University, December 1988).
5. See Stephen Henningham, *The Pacific Island states: Security and Sovereignty in the Past Cold War World* (Londen; Macmillan Press Ltd, 1995).
6. David Hegarty "Small State Security in the South Pacific", *Security of Small States*, p.158.
7. Ken Ross "South Pacific in Contemporary Political and Security Developments", *The Round Table*, vol.342, 1997, p.269.
8. For details see R.A Herr, "Regional Security in the South Pacific" *Pacific Islands Year Book*, 13th edition (Sydney: Pacific Publications, 1978) pp21-31.
9. See Te'O I.J. Fairbairn & other, "The Islands and the World" in *The Pacific Islands: Politics, Economics and International Relations, East West Centre International Relations Program*.
10. Ibid.
11. Amba Pande, "Race and Power Struggle in Fiji", *Strategic Analysis*, Vol.24, No.6, pp.1155-1169.

12. Man Mohini Kaul, *Pearls in the Ocean* (New Delhi: UBS Publishers, 1993), pp.51-64.
13. Ibid., pp.64-74.
14. See, Stephnene Hennigham, n.5, pp.71-90.
15. See for details Anthony Bergin, "Fisheries and the South Pacific", *Asia Pacific Community*, No.22, Fall 1983, pp.20-32; and W.M. Sutherland, "South Pacific Forum Fisheries Agency", *Ocean Development & International Law*, Vol.18, No.6, p.1987.
16. See, for example, Norman Meller, "The Pacific Islands Microstates", *Journal of International Affairs*, Vol.41, No.1, Summer/Fall 1987, pp.122-27. Also Jim Sandy, "South Pacific Culture Politics", *Working Paper No.174*, SDSC, ANU, Canberra (December 1988), pp.10-12.

9

Nuclear Activities

NUCLEAR TESTING

The South Pacific region became an attractive site for the great powers to test their nuclear devices, because of its remoteness and relative isolation. Americans were the first to start using the Pacific for the nuclear tests. They dominated the Micronesian Islands of the western and southern Pacific and conducted the first test at Bikini Atoll in July 1946 in the full glare of publicity. The operation was called 'operation cross roads' and the first test was known as 'Able'. The second test was called 'Baker' and was also conducted in the same month. Both the explosions created huge columns of cloud and water more than a kilometre high.[1] The scientists at Bikini clearly understood the dangers and by August 1946 recommendations, were made to prohibit the use any marine animal within 100 miles of Bikini. A cautious scientific survey of Bikini in 1947 showed that, while radioactive material had been flushed from the waters of the lagoon into the open sea, the mud at the bottom remained 'highly radioactive'.[2]

In all, the United States conducted 66 tests at different sites in the Pacific like Envitak and Marshall Islands, where the famous Hydrogen bomb 'Bravo' was exploded in 1954. There was a great deal of controversy over the effects of Bravo on Marshall Islanders. Several islands were destroyed and many others were so irradiated that they became uninhabitable.[3] Initially, the inhabitants agreed to vacate their homes on

American assurance that they would be able to return when the tests were over. But they were unable to do so due to the danger of radioactivity. It may be mentioned that the painful experience of Marshall Islanders had been one of the major causes which gave rise to the strong anti-nuclear sentiments of the Pacific people.[4]

The memory of the sufferings of Marshall Islanders were strengthened by a film 'Half life', widely circulated during 1986. The film made by Darnis O'Rourke, suggested that the United States allowed Marshall Islanders to die of a scientific test of radiation.[5]

Another country which was indulged in nuclear testing in the South Pacific was Great Britain. Britain conducted a series of nuclear tests in Australia between 1952 and 1957 and on Christmas Islands between 1957 and 1958. The Australian government allowed the British to use its sites at Monte Bello, Emu and Maralinga, boasting of a 'striking example of inter commonwealth cooperation'.[6]

Thus, for the sake of Britain whose nuclear weapons did nothing for Australia's defence, Australia willingly accepted radioactive contamination. England had the bomb and; Australia had the fallout.[7] However, after the signing of Partial Test Ban Treaty in 1963 both the United States and Britain stopped using the Pacific sites for nuclear tests.[8]

France, which was considered as the main culprit, began atmospheric testings at Mururoa Attol in 1966, after it was forced to abandon the Sahara test site following the Algerian independence in 1962. It established a 'Centre d'experimentation du Pacific' at Mururoa Atoll in Tuamotu. The French dominated the Polynesian group of Islands. Since 1966 France continued with its nuclear tests, with few gaps in between till late 1990s and attracted strong opposition from the regional states. This opposition was intensified after the election of labour governments in Australia and New Zealand. Later, Australia, New Zealand and Fiji challenged the French nuclear testings in the International Court of Justice and other world fora. The effect of this regional opposition, together with the weight of the world opinion made the French shift from atmospheric testing to underground testings from 1975 onwards.[9]

On its part, France justified its nuclear testings on three grounds: first, it argued that France's right to carry out its tests and to maintain its security should be recognised as had been in the case of Britain, the United States, former USSR and China; second, that under French law all the overseas French controlled territories are considered as being a part of France and hence, the tests were being undertaken on the French soil and that therefore, no one had the right to raise objections, and finally, the French government contended that the tests were safe.[10]

France even invited research groups to test the pollution level at the Muroroa atoll. The first of these missions was the 'Atkinson Mission' (1983) which had members from Australia, New Zealand and Papua New Guinea. The mission made its research in guided circumstances and although expressed general concern over the testings, refuted the claims made by anti-nuclear groups about health and environment.[11] But concerns persisted and island states showed no sympathy for the French arguments. The general mood was reflected in Vanuatu's Prime Minister Father Walter Lini's statement, "our position still is, that we oppose the French Nuclear tests on moral grounds and not on the premise that the tests are scientifically dangerous.... Obviously, if the French government thinks that the tests are safe, why does it not experiment with them in France? The fact that the French government continues to carry out its nuclear tests in the South Pacific, thousands of kilometres away from France, is itself morally wrong and no scientist can convince us otherwise."[12]

France, however, continued with its tests without paying any heed to the regional protests and even resorted to terrorist kind of activities in the region. The French agents sabotaged a protest ship 'Rainbow Warriors' belonging to an international environmental group 'Green Peace' on 10th July 1985 on Auckland Harbour. It caused international furore and diplomatic embarrassment for France. At first, the French government denied its involvement in the affair. However, later on it was proved that France was involved and the French President Mitterand had to order an enquiry into it. The Rainbow Warrior disaster was taken very seriously the world over and France's political rating fell even lower.[13] However, even after this incident France continued with its nuclear testings till the late 1990s.

NUCLEAR DUMPING

Apart from nuclear testing, another major long-term threat that the people of the South Pacific had been facing is the dumping of nuclear wastes. There was a strong opposition to the Japanese Plan to dump upto 10,000 barrels of radioactive waste north of the Mariana Islands during 1980s.[14] The United States was already using the California coast line to dump nuclear wastes. This issue was taken up by the South Pacific Forum in its subsequent meetings and finally, Japan had to abandon its plan of dumping nuclear wastes in the Pacific. The decision to prohibit the dumping of nuclear wastes in the area was included in the Treaty of Rarotonga. This reflects the concern of the South Pacific Forum that the region should not become a dumping ground of nuclear wastes. But economic constraints tempted these island countries to accept offers for dumping of nuclear wastes in return of money. Several of them like Marshal islands, Western Samoa, Tongo and Papua New Guinea were presented with such offers by the affluent countries.[15]

OTHER NUCLEAR ACTIVITIES

There were also certain other forms of nuclear activities in the region which became an issue of concern for the countries of the

South Pacific. For example, the Americans and French brought to the area the most hi-tech nuclear weaponry. The United States also started deploying Trident submarines each equipped with 24 Trident and SLBMs (Submarine Launched Ballistic Missiles) in the Indian and Pacific Ocean.[16] Moreover, there is the problem of military bases. Although the US follows the policy of non-discloser but to all the probability the bases have nuclear capability. There are mainly three American military bases which come within the zone of the Treaty of Rarotonga and these are in Australia. The United States installations at North-West Cape, Pine Gap and Nurrungar are central to the American interest in the region and also for the Australian-American relationship.

North-West Cape

It is presently one of the most important links in the United States global defence network. Its main purpose is to maintain reliable communications with the submarines of the United States fleet serving in the area and in particular, to provide communication for American navy's most powerful deterrent force—the nuclear powered ballistic missile submarines.[17]

Nurrungar

It is located in the Woomera restricted area about 480 kilometres north-west of Adelaide. It was one of the two ground stations for the Amercan satellite early warning system. Nurrungar provides a real time data link between the North American Air Defence Command (NORAD), the Strategic Air Command (SAC), and the National Military Command system on the one hand, and the satellite early warning system on the other.[18]

Pine Gap

The Pine Gap facility which became operational in 1969, is located 19 kilometres south-west of Alice Springs. "The 'business end' of the facility currently consists of seven large radomes and enormous computer complex". Pine Gap is controlled by the CIA with extensive participation of NASA.[19]

These installations and facilities on the Australian soil are perceived by the people of the South Pacific as possible targets of attack posing nuclear threat not only to Australia but to the entire region.[20] Other than the nuclear threat there were some other factors too which prompted the idea of establishing Nuclear Weapon Free Zone in the South Pacific.

REFERENCES

1. For details, see Stewart Firth, *Nuclear Playground* (Sydney: Allen & Unwin, 1987).
2. Ibid., p.7.
3. Editorial, *Bulletin of Concerned Asian Scholars*, Vol.18, No.2, April-June 1986, p.9.
4. T.V. Paul, "Nuclear Free Zone in the South Pacific", *Round Table No.299*, July 1986, pp.90-91.
5. For details see Robert Milliken, "Half Life: Deadly Show", *Bulletin of Concerned Asian Scholars*, Vol.18, No.2, April-June 1986, pp.90-91.
6. *The Report of Royal Commission into British Nuclear Tests in Australia Canberra*, Australian Government Publishing Service, 1985, p.405.
7. Stewart Firth, n.1, p.10.
8. "South Pacific: A Not So Nuclear Free Zone", *Peace Studies* (Melbourne), October 1985, pp.6-8.
9. Kevin C. Clements, "New Zealand's role in promoting Nuclear Free Pacific", *Journal of Peace Research*, Vol.25, No.4, December 1988, pp.395-410.
10. Greg Fry, "A Nuclear Free Zone for the South Pacific: Prospects and Significance", *Working Paper No.75* (Canberra: Australian National University, 1983), p.13.
11. Report of a New Zealand, Australian, Papua, New Guinea Scientific Mission to Mururoa Atoll, New Zealand, Ministry of Foreign Affairs (Wellington, 1984).
12. Interview with David Robie, *Vanuatu Weekly*, 25 August 1984, pp.15-16.
13. For details, see Ramesh Thakur, "A Dispute of Many Colours: France, New Zealand and Rainbow Warrior Affairs", *World Today*, Vol.92, No.12, December 1986, pp.209-14.
14. See South Pacific forum Communiques, 1971-85, reported in *Australian Foreign Affairs Record*, 1971-85.
15. *Canberra Times*, 7-14 May 1990.

16. *Bulletin of Concerned Asian Scholar*, n.3, p.40.
17. Desmond Ball, "United States Installations in Australia", *Working Paper No.36*, SDSC, Australian National University, Canberra, June 1981, p.3.
18. Ibid., p.4.
19. Ibid., pp.4-5.
20. Ibid., pp.7-8.

10

Other Nuclear Concerns

THE PRESENCE OF SOVIET UNION AND THE AMERICAN ATTITUDE TOWARDS SOUTH PACIFIC

According to Ramesh Thakur, a noted scholar on the area, "the two factors which accelerated the process of declaration of the South Pacific as nuclear free zone during 1980s were the growing presence of Soviet Union and the nuclear policies of the Reagan administration."[1] Some other scholars also are of the opinion that the Soviet posture and policy in the Pacific enhanced the nuclear peril. The danger they thought stemmed from the Soviet weakness in conventional force in the region which prompted it to rely heavily on nuclear weapons. Hence, the Pacific, they thought was deeply and dangerously entangled with the superpower nuclear arms race.[2] Former Commander of the Pacific Command Admiral Robert Leng, went on record, as saying that the Pacific is the region where we shall witness confrontation with the Soviet Union.[3] The Australian Prime Minister Malcom Fraser also said in Washington in 1985 that the Soviets might get some type of military base in the Pacific very soon and they would be 'just an hour by jet' from Hawaii.[4]

Going by these arguments a new dimension may be added to the need for establishment of nuclear weapon free zone in the South Pacific. As the war was evident and the possibility of nuclear attack appeared to be inevitable hence, there was a need for the nuclear weapon free zone. But as one analyses these arguments, it appears that in order to justify the United

States presence in the South Pacific, the pro-defence groups grossly over-exaggerated the "Soviet threat".

As far as the United States presence in the Pacific was concerned it had been the most important external power present in the region. The South Pacific has for long revolved around the ANZUS security umbrella to the extent that it has come to be known as the 'ANZUS Lake'. During the World War II, the United States had left a very favourable impression in the minds of the Pacific Islanders and had enjoyed high degree of popularity. But by the end of 1970s and 1980s, the island states became largely concerned with the United States stand on tuna fishing (as has been already discussed) and environmental hazards caused by the United States nuclear tests. All this led to a sharp decline in the US popularity.[5]

As for the United States, the region's importance was basically strategic,[6] but over all, the South Pacific was generally low on the US agenda[7] as against the western and northern Pacific. Although Marshall Islands in the South Pacific possessed the Kwajalein missile range facility which was essential for ICBM testing and development of Strategic Defence Initiative, but most of its 'fall back' sites like Guam, Palau and Northern Mariana Islands remained in the western and northern Pacific. Nevertheless, the continuing strategic relevance was evident in the 1951 establishment of ANZUS alliance.[8]

However, a major turning point in the United States attitude (along with Australia and New Zealand) could be seen after 1976 when Soviet Union began its efforts to develop relations with the region. In 1976, the Soviets offered economic assistance to Tonga in return of fishing facilities. It triggered a major reassessment on part of ANZUS partners. As a consequence, there was an expansion of Australia and New Zealand's regional diplomatic presence and developmental assistance programmes and also there was a pressure on Washington to play a more active secondary role in the region.[9] Thus, a new policy was adopted by the ANZUS partners for dealing with the South Pacific which came to be known as the policy of 'strategic denial'—a policy concerned with denying any Soviet influence.[10] The policy meant attempting to exclude any Soviet economic or diplomatic involvement in the region on the ground that it could be a stepping stone to a military base. As more Soviet initiatives came up, the policy of 'strategic denial' was followed more vigorously. With the advent of Mikhail Gorbachev a new colour was added to the Soviet Union's Asia Pacific Policy.

The change in the Soviet policy resulted in an increased Soviet attention which was manifested through fishing deals with Kiribati in 1985 and Vanuatu in 1987. The United States Department of Defence reacted with exaggerated alarm over these two incidents. However, the fisheries' agreements were not renewed after a year because of Soviet insistence on lower fees.

The commercial component appeared to be more important for Soviet Union than the political profile in the region stemming from fishing activities and after that no major initiative came up from the Soviet side.[11] Thus, it appears that the 'Soviet are coming' syndrome was projected as an overrated threat by the ANZUS partners but if not anything else the so-called threat certainly managed to increase the US interest in the region.

CONDUCIVE REGIONAL ENVIRONMENT

Another point highlighted by Ramesh Thakur is that there were some positive elements within the region which favoured the establishment of nuclear weapon free zone, they were the general peace and harmony prevailing in the region. There was no armed conflict between states in the region or fear of war with states outside the region. In case of any conflict situation the regional countries may want to keep open the option to go nuclear.[12] This point has also been highlighted by the United Nations as has been discussed in Chapter 1.

Another noted scholar Greg Fry who has written at length on this issue, believes that the fate of any nuclear free zone proposal largely depends on the strategic situation within the region concerned. If there is any armed conflict among states in the region, or if states in the region are at war or fear of war with states outside the region, this would seriously affect the chances of gaining agreement on a nuclear free zone.

In the South Pacific far from there being any serious conflict, there has actually been a considerable degree of cooperation among the sates. This cooperation is highly institutionalised. There is a vast array of regional organisations dealing with wide range of areas such as fishing, economic development, health, trade, etc. There are regular and close consultation among the leaders of these states.[13] The nuclear issue has been one such area where form the very beginning alignment has exited among the forum Island countries which gave necessary fillip to the so called 'anti-nuclear movement' in the region.

RISE OF ANTI-NUCLEAR MOVEMENT

Almost for 40 years nuclear weapon states used the South Pacific as their playground and the Pacific Islanders have had enough. The Forum Island Countries, no longer under the colonial rule were free to think and act in their own way. They uniformly opposed the nuclear tests to the extent that the issue completely transformed the political landscape of the South Pacific.

The origin of the movement for a nuclear free and independent Pacific can be traced to the ATOM committee (Against Tests on Mururoa) which was formed in Fiji in 1970. It consisted of Pacific Theological College, the University of South Pacific and the Fiji YWCA. The first conference on nuclear free pacific was held under its auspices in 1975.[14] Several other conferences were held after this. Trade unions, private aid organisations, environmentalists, disarmament lobbies, women's collectives and Christian groups, all seemed to have taken over by anti-nuclear movement. And somehow the movement did manage to enjoy some successes in spite of its limited resources as compared to the might of the nuclear powers. Its lobbying helped to dissuade the Japanese from dumping radioactive wastes in the north Pacific.[15]

Another important conference in this regard was held in Vanuatu where a 'People's Charter for a Nuclear Free and Independent Pacific' was adopted. The Charter said, "our environment continues to be despoiled by foreign powers developing nuclear weapons for a strategy of warfare that has no winners, no liberators and imperils the survival of all human kind." It also said that military bases and testing facilities might have brought employment to the Pacific islanders but they had to pay a price of destruction of their custom, their way of life, pollution to their waters and a threat of disaster by radioactive poisoning.[16]

Thus, it is clear that the countries of South Pacific worked with excellent cooperation and were very well aware of the health and environmental hazards affecting them due to nuclear pollution. Several of the island countries like Solomon Islands and Tuvalu had banned nuclear armed ships into their wanters by mid-1980s when the nuclear issue reached its zenith.[17]

Although extremely short of resources and scientific infrastructure, they tried to manage the situation by uniting under the banner of the South Pacific Regional Environment Programme (SPREP) which was launched in 1982. The programme is based in Apia, Western Samoa.[18] The issue was also taken up in the South Pacific Forum time and again.[19] Other than the South Pacific Forum, Australia and New Zealand being two major countries of the region, have also played extremely important role in the anti-nuclear movement and establishment of South Pacific Nuclear Free Zone and in winning international support for it. In 1973 proceedings were instituted by Australia and New Zealand in the International Court of Justice (ICJ) in Hague against France for carrying on dangerous nuclear experiment affecting the health of the people of the South Pacific nations. Though the ICJ issued an interim injunction against France ordering it to cease the testings, France disregarded it. But in face of the worldwide protests France stopped atmospheric testings but continued with underground tests.[20] Nevertheless, as far as Australia and New Zealand were concerned, there were some basic differences in their nuclear stand and hence, it would be appropriate to discuss the differences between Australia and New Zealand separately.

NEW ZEALAND'S ANTI-NUCLEAR STAND AND THE COLLAPSE OF ANZUS

From the very beginning New Zealand took a lead in the anti-nuclear movement in the South Pacific. As far as the International nuclear arms control measures are concerned, New Zealand always has been an active player. It was among the founding signatories of Non-Proliferation Treaty (NPT) and was a constant supporter of comprehensive test ban treaty. It also played a very active role in generating international support for the nuclear issue in the South Pacific.

In 1975, New Zealand put forward the idea of establishing nuclear weapon free zone in the South Pacific in the South Pacific Forum. Thereafter along with Fiji, the New Zealand Government wrote to the United Nations Secretary General requesting him that the issue be included in the forthcoming 30th Session. Consequently, the draft resolution was adopted (co-sponsored by Fiji and Papua New Guinea) in the first committee on 28 November by a vote of 94-0, with 18 abstentions; and in the General Assembly on 11 December by a vote of 110-0-20 (Resolution 3477). China was the only nuclear power to vote in favour. The other four nuclear powers namely

Britain, France, USA and USSR expressed general sympathy, but abstained on vote because of fears for their rights upon the high seas.[21]

It was a major step towards internationalising the nuclear concerns of the Forum Island Countries. The issue was particularly taken up by the New Zealand's Labour Party and it could be pursued more vigorously after its coming to power in 1972.[22] In fact, nuclear concerns became one of the elements of New Zealand's foreign policy and it also gave hints that it could inhibit the operations of the United States Navy in the Pacific.[23] But this policy could not be pursued further as the Labour's tenure was short-lived and with the coming of the National Government under R.D. Muldoon the issue ceased to be important. One of the earliest moves of the National Government was to renounce the support the Labour's concept of nuclear weapon free zone in the South Pacific.[24]

But again with the return of the Labour government in 1984 under David Lange the issue came to fore. Defence and foreign policy issues were actively debated during the election campaign. The victorious Labour Party repeated its commitment to a policy of keeping all nuclear vessels out of New Zealand which meant New Zealand's rejection of the United States nuclear activities in the region. It also reflected nuclear pacifism on part of New Zealand, as it was willing to underplay its trade and military ties with the United States vis-a-vis a higher goal of de-nuclearisation of the South Pacific.[25]

As far as the United States was concerned, New Zealand's nuclear allergy was always a cause of concern for it. New Zealand's Labour Prime Minister Kirk had raised this issue at the 1974 meeting of the ANZUS council, in response to which the US Secretary of State had said that the proposed zone would pose very serious problems for the United States navy which was converting to a nuclear force.[26] The United States was also worried that this attitude could also encourage like-minded groups in other allied countries.[27] The labour government in New Zealand valued the American alliance and never intended to end the ANZUS relationship. In 1975 itself, New Zealand's Prime Minister had told Australia that he thought its relationship with the United States could accommodate the proposal without undue restrain.[28] With same belief New Zealand went ahead with its anti-nuclear campaign backed by a strong public opinion.

On 27 January 1985, the United States government asked for a conventionally powered ship 'US Buchanan' to visit New Zealand which was turned down by the Lange government. Although at the same time he also stated that New Zealand would remain committed to the ANZUS alliance. The United States Secretary of Defence Casper Weinberger characterised the New Zealand stand as a serious attack on the alliance and the US responded by suspending cooperation with New Zealand in training, military courses and exercises and sharply curtailing intelligence shared with New Zealand.[29]

As a result, the ANZUS broke down for the time being. David Lange himself wrote that the development was viewed with concern by governments in the region as they had, over the years remained under the United States security umbrella.[30] Nevertheless, he continued with his vigorous commitment to anti-nuclear policies and gave a fresh impetus to the proposal of South Pacific-nuclear-free zone.

AUSTRALIA'S SECURITY CONCERNS AND ESTABLISHMENT OF NUCLEAR WEAPON FREE ZONE

Australia, the third ANZUS ally had a precarious position as far as nuclear issue in the region was concerned. On one hand, its alliance with the United States was of utmost importance for its security, but on the other, it was also worried about the growing nuclear activities in the region. At the same time Ausatralia was also concerned about its 'big brother' image in the region and was very much sensitive towards the expectations of the Forum Island Countries.[31] However, it was the only country in the region which was involved in any kind of nuclear activity. As already discussed, it had three nuclear bases belonging to the United States on its soil. Other than the US bases, Australia was also exporting uranium to the countries like France, which had generated a serious public debate inside Australia. A decisive portion of public opinion favoured complete prohibition of uranium exports. But Australia had always argued that its uranium export are basically used for peaceful purposes under IAEA supervision.[32]

As far as the establishment of nuclear free zone in South Pacific is concerned, it was being debated in the Australian Labour Party for quite some time but Australia was hesitant about the proposal of banning nuclear powered ships into the pacific waters.[33]

However, at the same time it went ahead with supporting New Zealand in the South Pacific Forum. But in wake of the US concern about passage of its ships, port call and other nuclear activities in the region the Australian Prime Minister Witlam had written to New Zealand's Prime Minister, that nuclear free zone would be 'disadvantageous to the United States strategic interest and security arrangements in the South Pacific', specifically it would pose problems for ANZUS.[34]

However, because of the change of governments in Australia and New Zealand the issue became irrelevant. But the 1975 episode made one thing clear that transit of nuclear armed ship, and probably the nuclear bases were the issues which provoked the United States opposition.

With the coming of Labour government in Australia under Robert Hawke the issue once again came to forefront. He came up with the proposal of South Pacific Nuclear Free Zone Treaty (SPNFZT) in the South Pacific Forum meeting in 1983. The proposal naturally took care of the United States concerns regarding transit and other issue and evoked different reactions from the regional powers. It was felt by some analysts that Australia took initiative due to the fact that David Lange was about to come up with a proposal which could have led to the banning of United States nuclear ships in the South Pacific.[35] The regional countries like Papua New Guinea, the Solomon Islands and Vanuatu wanted a more radical treaty while Fiji, Western Samoa, Tonga, Cook Island and Nive wanted to act cautiously even in relation to this modest treaty.[36] However, the Treaty of Rarotonga, as it exists today is based on the proposal made by Australia.

REFERENCES

1. Based on a personal interview with Ramesh Thakur on 21 January 1994.
2. See, for example, Peter hayes and Lyuba Zarsky, "Pacific Arms Control and Regional Initiatives", in Rangihui Walker and William Sulherland, (ed.), *The Pacific: Peace, Security & the Nuclear Issue* (Tokyo: The United Nations University, 1988), p.74.
3. Peter D. Jones, "Militarisation of the Pacific" in ibid., p.67.
4. The Australian, 24 April 1985.
5. See Stephen Henningham The Pacific Island States: Security and Sovereignty in the Post Cold War World (London: Macmillan Press Ltd., 1995), p. 93. and John Dorrance, "Soviet Union and the Pacific Islands: An American Assessment and Proposed Western Strategy", *Working Paper,* No.207, ANU, Canberra, May 1960, p.9.
6. John C. Dorrance, *The United States and the Pacific Islands* (West Port Conn. And London, Prager, 1992), pp.60-68.
7. Keith D. Suter, "The United States and the South Pacific: A Benign Neglect", *Contemporary Review,* Vol.258, No.1503, April 1991, pp.
8. John C. Dorrance, "The Soviet Union and the Pacific Islands: An American Assessment and Proposed Western Strategy?", *Working Paper,* No.207, ISDS, ANU, p.1.
9. Ibid., p.6.
10. See R.A. Herr, "Regionalism, Strategic Denial and South Pacific Security", *Journal of Pacific History*, No.21, October 1986.
11. Ramesh Thakur, *The South Pacific: Problems, Issues and Prospects* (Otago: Macmillan, 1991), p.19.
12. Excerpts of the interview with Ramesh Thakur in 1994.

13. See Greg Fry, *Working Paper,* No.75, SDSC, ANU, p.44.
14. Vijay Naidu, "The Fiji Anti-Nuclear Movement: Problems and Prospects", Paper to the UN University Conference, Auckland, April 1986, p.5.
15. Stewart Firt, *Nuclear Playground* (Sydney: Allen & Unwin, 1987) p. 134.
16. Nuclear Free and Independent Pacific Conference—1983, 10-20 July 1983, Port Vila, Vanuatu, p.27.
17. Paul F. Gardner, "Tuna Poaching and Nuclear Testing in the South Pacific", *Orbis*, Vol.32, No.2, Spring 1988, p.258.
18. For details, see, Bryant J. Allen, "The Pacific Islands: Background of the Region", *Europa Yearbook*, 2001, pp.882-887.
19. See http:#chacmool.sdnp.undp.org/pacific/forumsec/Docs.
20. Manmohini Kaul, Pearls in the Ocean (New Delhi: UBSP, 1993) p. 82.
21. Ramesh Thakur, "A Nuclear-Weapon Free South Pacific: A New Zealand Perspective", *Pacific Affairs*, Vol. 58, No. 2, pp.218-219.
22. For details see David Lange, *Nuclear Free: The New Zealand Way* (Penguin Books).
23. David J. MCcrow, "New Zealand's Foreign Policy under National and Labour Governments: Variations on the Small State Theme?", *Pacific Affairs*, Vol.55, No.4, Winter 1992-93, p.18.
24. *New Zealand Herald*, 21 January 1976, p.1.
25. *New Zealand Official Yearbook*, 1986-87, p.577.
26. Margaret Hayward, *Diary of Kirk Years* (Queen Charlotte Sound and Wellingon: Cape Catley and AH and AW Reed, 1981), p.108.
27. *The Washington Post*, 15 February 1985.
28. *The Nation* (Wellington), No.3, 20 November 1980, p.7.
29. Dora Alves, "United States-New Zealand Relations: The National Government of New Zealand", *Asian Survey*, Vol.31, No.11, November 1991, p.1065.
30. David Lange, "New Zealand's Security Policy", *Foreign Affairs*, Vol.63, No. 5, Summer 1985, 00.1009-1019.
31. See Amba Pande, "Australia's Policy Towards South Pacific", in D. Gopal, (ed.), *Australia in the Emerging World Order* (Delhi,: Shirpa Publications, 2002, pp.189-198.
32. http:#www.energyres.conn.au/uranium or http: # www. nuclear-disarmament. org /polcies.com.
33. *The Nation*, (Wellington) No.3, 20 November, 1980, p.7

34. As reported in J.R. Walsh and G.J. Muster, (ed.), *Documents on Australian Defence and Foreign Policy*, 1968-1975, (Canberra, 1980).

35. Greg Fry, "The South Pacific Nuclear Free Zone", *SIPRI Yearbook*, 1986, p.501.

36. Man Mohini Kaul, No.20, p.84.

11

Assessment of the Treaty of Rarotonga

ORIGIN OF THE TREATY

The concept of nuclear weapon-free zone caught the attention of the Australian Labour Party (ALP) as early as 1962, after successful denuclearisation of Antarctica in 1959, and the New Zealand Labour Party was drawn to it after the successful completion of Tlatelolco Treaty in 1959.[1] Since then the Labour parties of both the countries pursued the idea, and as already discussed, the idea was put forth in the South Pacific Forum meeting in 1975 and was commended by the member countries. Later on, it was even adopted as a UN General Assembly resolution. However, the idea of establishing a nuclear weapon free zone in South Pacific could not take a practical shape. This was partly, because of the lack of enthusiasm on Australia's part in view of its security concerns and also because of the change of governments in both the countries. With the coming of conservative governments to power the proposal got completely shelved. The subsequent South Pacific Forum meetings, held after 1976, dropped the idea from their agenda although the anti-nuclear movement continued to grow. The issue was again revived in 1983 by the Australian leadership.

The return of Labour governments in Australia and New Zealand in the early 1980s provided the requisite impetus to the proposal for Nuclear Free Zone (NFZ), in South Pacific. In the

South Pacific Forum meeting held in Canberra on 28-29 August 1983, the Labour Prime Minister Robert Hawke put forward the proposal. The Australian proposal took care of the US security concerns and was mainly aimed at banning the French nuclear testings in the region. The basic aim was to put the spotlight on France.[2] As a result, it did not create any undue concern on the part of the United States. However, the proposal was referred back for further consultations[3] because of some differences within the Forum members as discussed earlier. But in spite of the differences the proposal was endorsed in the 1984 South Pacific Forum meeting held in Tuvalu and a working group was set up to draft the treaty.[4]

The new Labour government of New Zealand gave further boost to the proposal and finally, in the 16th South Pacific Forum meeting held in August 1984 held in Rarotonga on 4-6 August 1985, the South Pacific-Nuclear-Free Zone Treaty was approved and signed.

THE TREATY AND ITS PROVISIONS

The treaty came into force in December 1986 and its provisions are:

Name of the Treaty

The official title of the Treaty of Rarotonga is South Pacific Nuclear Free Zone Treaty and not South Pacific Nuclear Weapon Free Zone Treaty. However, both these terms are not entirely correct. The fact was that the nuclear weapons continued to be present through transit and visits of nuclear armed ships and aircraft. However, at the same time the treaty does not prohibit peaceful nuclear activities in the region. Thus, as Paul F. Power points out "the adopted title gives a misleading impression".[5]

Geographical Scope

Article 1 of the treaty states, "South Pacific Nuclear Free Zone" means the areas described in annexure 1 as illustrated by the map attached to that annexure.[6] Annexure 1 of the treaty describes in detail the geographical limits of the treaty. Roughly the area the treaty covers is as follows: in the north the boundary is the equator except for certain areas north of it (which include Exclusive Economic Zones of Papua New Guinea. Nauru and Kiribati); in the south, the northern boundary of demilitarised Antarctic area; in the east, the western boundary of the Latin American free zone; and in the west, the north-south extension of the western coast of Australia, the Australian coast and the frontier between it and Irian Jaya. Apart from this, article 12 of the treaty states that "if a member of the South Pacific Forum whose territory is outside the South Pacific Nuclear-Free Zone Treaty becomes a party to the treaty, Annex I shall be deemed to be amended so far as is required to include at least the territory of that Party."

Renunciation of the Nuclear Explosive Devices

Under the treaty the parties undertake, "not to manufacture or otherwise acquire, possess or have control over any nuclear explosive device by any means anywhere inside or outside the region, nor to seek or receive any assistance for such purposes, nor to assist or encourage any state in such activities." There is a clear definition of nuclear explosive device in Article 1(c) of the treaty which states "any nuclear weapon or explosive device capable of releasing nuclear energy, irrespective of the purpose for which it could be used. The term includes such a weapon or device in unassembled and partly assembled form." In this respect the South Pacific Nuclear Free Zone Treaty goes one step ahead of the Non-Proliferation Treaty (NPT) and the Treaty of Tlatelolco.

Peaceful Nuclear Activities

About the peaceful use of nuclear energy, the Rarotonga Treaty states, in Article 4, "not to provide source or special fissionable material or equipment or material specially designed or proposed for the processing, use of production of special fissionable material for peaceful purpose" to any non-nuclear weapon state unless subject to safeguard required, under the provisions of NPT or IAEA. This Article underwent a great controversy as Australia was exporting uranium. Obviously, in the event of a prohibition, the exports could have had negative consequences.[7]

Prevention of Stationing of Nuclear Explosive Device

Article 5 of the treaty has two paragraphs. The first paragraph deals with stationing of nuclear explosive devices and states, "Each Party undertakes to prevent in its territory the stationing of any nuclear explosive device." The term stationing has been described in Article 1(d) of the treaty in detail: "Stationing means emplantation, emplacement, transportation on land or island waters, stockpiling, storage, installation and deployment."

The second paragraph of Article 5 of the treaty describes the right of every state party to the nuclear-free zone to decide for itself whether to allow visits by foreign ships and aircraft to its ports and airfields; transit of its airspace by foreign aircraft and navigation by foreign ships in its territorial sea or archipelagic sea lane passage or transit passage of straits. This clause clearly shows that visit or transit by nuclear ships is permissible. This issue was the most controversial one.

There is also a legal angle to it as discussed in Chapter 1. The question is whether or not to respect the freedom of navigation into the high seas. In the case of South Pacific since the territorial lands are typically specks in an ocean vastness, this clause becomes all the more significant. Moreover, the issue of transit of American ships was important for the regional security.

Prevention of Nuclear Tests

Article 6 of the Rarotonga Treaty forbids the state parties (a) "to test any nuclear explosive device", and (b) "to assist or encourage the testing of any nuclear explosive device by any state".

As Article 3 also forbids the parties to develop any nuclear explosive device the important point here is that it has nothing to do with the territories of member countries but with the nuclear weapon powers as they are the real threat. This matter specifically has been dealt with in Protocol 3 of the Treaty.

Dumping of Nuclear Wastes

Under the provisions of Article 7, the parties undertake:

a) Not to dump radioactive wastes and other radioactive matters anywhere within the zone;

b) To prevent such a dumping by anybody in their territorial sea;

c) Not to assist or encourage dumping; and

d) To support the conclusion as soon as possible, of the proposed convention relating to the protection of natural resource and environment of the South Pacific region and its protocol for the prevention of pollution of the South Pacific by dumping. The second paragraph of this article adds that the undertaking as mentioned above in (a) and (b) shall not apply to the countries that ratify the future convention and protocol. Dumping of nuclear wastes had been one of the major factors which initiated the signing of the Treaty of Rarotonga.

Control System

The control system established by the Rarotonga Treaty includes the application of IAEA safeguards (Article 8). Annexure 2 of the Treaty deals with this subject. There is also a provision of the controlling body, the 'Consultative Committee' with the power to direct a special inspection team to investigate any suspected violation on the territory of a member state. But there is no sanction against non-compliance provided for in the treaty.

As in the case of most of the arms control agreements, the ultimate sanction is the breakdown of the agreement itself. However, since the violation means violating the arms control system in general and also the international law, it acts as a sanction. Another factor in this regard is that in the case of the South Pacific, since all the states closely cooperate in the field of security, economy, etc., any violation of the treaty may jeopardise other objectives also which act as an added obstacle. The treaty also establishes a 'complaints procedure' in Annexure 4.

Reports and Exchange of Information

The Treaty of Rarotonga does not establish a new body for the submission of report by member countries, as it was done in the treaty of Tlalelolco. Article 9 of the treaty states, "Each party shall report to the Director of the South Pacific Bureau for Economic Cooperation, as soon as possible, about any significant event within its jurisdiction affecting the implementation of this Treaty.

The Director shall circulate such reports promptly to all parties (Article 9). The article further states "the parties shall endeavour to keep each other informed on matters arising under or in relation to this Treaty," and, "the director shall report annually to the South Pacific Forum on the status of this Treaty."

Consultations and Review

Article 10 of the Treaty states that, "without prejudice to the conduct of consultations among parties by other means, the director, at the request of any party, shall convene a meeting of the Consultative Committee, established by Annexure 3, for consultation and cooperation or any matter arising in relation to this Treaty or for reviewing its operation."

Amendment

Article 11 deals with the amendment procedure of the treaty. When a proposal is made by one of the parties, it is the duty of the director to circulate it among the other parties, not less than three months, prior to the convening of the Consultative Committee for this purpose. Any proposal agreed upon by consensus of the Consultative Committee shall be circulated for the acceptance of all parties. An amendment will enter into force 30 days after receipt by the depository of acceptance from all parties.

Signature and Ratification

This aspect is dealt with in Article 12 of the Rarotonga Treaty. It states that the "Treaty shall be open for signature by any member of the South Pacific Forum." Another paragraph in the same Article says, "if a member of South Pacific Forum whose territory is outside the South Pacific nuclear free zone treaty becomes a party to the treaty, Annexure I shall be deemed to be amended." In that case, if any of the countries such as Federated States of Micronesia, the Republic of Palau or the former United States Trust Territories tries to become a party to the treaty, the process will be simple and short.

Withdrawal

The right of withdrawal has been given in Article 13 of the treaty. It states "the treaty is of permanent nature and shall remain in force indefinitely." It further says, "withdrawal shall be affected by giving notice twelve months in advance to the director who shall circulate such notice to other parties." However, the article also makes it clear that withdrawal is possible only in the event of violation by any party of a provision of the treaty.

Final Clauses

Articles 14, 15 and 16 of the treaty deal with reservation, entry into force, and depository functions respectively. Apart from these provisions the treaty had three protocols which were to be signed by the nuclear powers.

The Protocols

Under Protocol 1, three expected signatories—France, the United Kingdom and the United States—undertake to apply a number of key treaty provisions to their territories in the region.

Protocol 2 is open for signatures by all the five nuclear-weapon states and asks them to observe the Treaty and refrain from using or threatening to use any nuclear explosive device against the parties to the treaty or such territories as are incorporated in the treaty.

Protocol 3, addressed to all the five nuclear weapon states, requests a promise not to conduct any test of a nuclear explosive device within the SPNFZ.

The three protocols were not open for signature at Rarotonga, together with the treaty. It was decided the time that before doing so, consultations with the concerned states should be held and finally, in the 17th Forum meeting held in Suva in 1986 it was decided that protocols should be open for signatures in December 1986, when the treaty enters into force.[8]

Although the regional countries were united on the basic issue, i.e., to rid the region from nuclear menace, there were some differences among them over some of the provisions of the treaty. It may be noted in this regard that Vanuatu which was a left neutralist country and, had recognised Cuba and joined non-aligned movement, had barred visits of the United States since 1986 [9]. It had a totally different view regarding the treaty as against Tonga which was formally named as a 'friendly country' by the United States and did not approve of any anti-nuclear initiative such as a nuclear-free zone. Kiribati, New Zealand, the Solomon Island and Papua New Guinea had radical views about the transit of the nuclear-powered ships while the countries like Australia and Fiji were towards the middle.[10] Nevertheless, all the South Pacific Forum countries signed the treaty by 1987 except for Tonga which signed it as late as 1996.

ATTITUDE OF NUCLEAR WEAPON STATES

Approval of regional countries was necessary for the treaty's coming into force but support of the nuclear weapon states was critical for its effective implementation. Hence, it is important to discuss their reactions.

As far as the United States is concerned, it refused to sign the protocols at the initial level, although, its security interests had very clearly been taken into consideration in the treaty. It was also advisable for the United States to do so in view of the growing anti-US and anti-nuclear feelings. As Ramesh Thakur opined, "if the treaty helps to appease the strong anti-nuclear sentiments in the region without undermining the United States, strategic position, then Washington would be better advised to settle for the moderate, Australian initiated zone than to inflame anti-Americanism by unilateral dissociation."[11] But the United States declared in 1987 that in view of its global security interests and responsibilities, it was not "under current circumstances" in a position to sign the protocols.[12]

This decision was all the more unfortunate because the United States had ratified both the protocols of the Tlatelolco Treaty. But probably the United States concern was that by signing the protocols it would be endorsing New Zealand's ban on nuclear ships. The United States Deputy Assistant Secretary of East Asian and Pacific Affairs stated that "there is a feeling that it would encourage other areas to adopt similar or more radical nuclear-free zones . . . . Another argument is that over a period of time perhaps even the South Pacific Nuclear-free Zone would be tightened."[13]

Another United States concern was its relations with France. A senior United States Defence Department official was quoted as saying that 'the strong opposition of France was the major element in the United States decision' since French testings make them able to participate as part of the deterrent of the West against the Soviet bloc.[14] The United States had also maintained a diplomatic silence over the French nuclear tests and the issue of New Caledonia much to the displeasure of the Forum Island Countries. In spite of its different stand finally, the United States did sign the treaty along with France on 25 March 1996.[15]

The United Kingdom, a close United States ally, followed the line of its partner by initially refusing to sign the protocols. However, it clarified that it respected the intentions laid out by Protocol I. Further, it also stated that it had no desire to conduct nuclear tests in the region, but still would like to keep the options open, "we intend to keep our attitude to the treaty and the Protocols under review."[16] In a news release the Australian Foreign Minister, Bill Hayden, expressed his concern over the British decision stating that it was difficult to understand the British decision.[17]

France, the main culprit, chose to overlook the treaty for the time being. It was the only nuclear power against whom the treaty was mainly directed to. But the Mitterand government thought that yielding to regional pressures would be a defeatist step. Shortly before the signing of the treaty the French government recommitted itself to continue its nuclear tests at Mururoa and after a brief moratorium it again started testings in 1995. "France defies the world, conducts third N-tests."[18] "Arrogant France conducts third nuclear tests."[19] Such were the captions used for France throughout the world. Sharp criticism came from Australia and New Zealand. "All he (Chirac) is doing, I think, is living in a sort of after-glow of the twilight of the cold war, still testing weapons," said the Australian Prime Minister Paul Keating.[20]

Defence analysts questioned whether France, after testing some 200 nuclear devices, needed any further tests. Strong feelings were also expressed in Germany and Italy.[21] Japan reacted in a blunt diplomatic language, "it is deeply regrettable that France defies protests from Japan."[22]

France however, also announced that it would sign the comprehensive Test Ban Treaty in 1996 after one final series of up to eight nuclear blasts[23] and also showed its readiness to join Pacific Non-nuclear Pact.[24] And when in January 1996 France announced a definitive end to its nuclear tests it was welcomed by the whole world. Finally on 26 March 1996, France signed all the three protocols of the South Pacific Nuclear Free Zone Treaty along with the United States and Britain putting an end to a 'sorry chapter' in the South Pacific history.

The most positive response for Nuclear Free Zone Treaty in South Pacific came from erstwhile USSR and China. Of course, Moscow had its own reasons for doing so. First, it desired to see an end to the United States security links with Australia and New Zealand and the wreck of ANZUS.[25] Second, it was also interested to see an end to the transit of the US nuclear ships and submarines following the lead taken by New Zealand in this context.[26] Third, the Soviet Union was also looking forward to improve its relations with the South Pacific countries. The Soviet attitude towards the treaty was welcomed by the regional countries. The Australian Foreign Minister welcomed Soviet Union's adherence to Protocol 2 and 3 of the treaty.[27]

China also responded positively to the Treaty of Rarotonga. It claimed to have supported the question of a nuclear-free zone as early as 1975 in the UN General Assembly. It signed Protocols 2 and 3 of the treaty which was welcomed by Australia and other countries of the region.[28]

ASSESSMENT OF THE TREATY

The Treaty of Rarotonga has indeed been a significant achievement for the South Pacific Forum and is one of the region's major contributions to global peace and security. It represented a genuinely collective effort by regional countries to express a deeply felt concern about nuclear testing and dumping of nuclear wastes. But after its coming into existence critics have come down heavily on the treaty. It has been described in several ways as 'a sham', 'a joke', 'a farce' or 'clay-tone zone'—a zone which is not a zone. It is said that the commitment by the 13 nations not to manufacture nuclear weapons is hardly of any importance as most of these nations, including Australia and New Zealand, are already parties to NPT. Moreover, these nations have no ambition to go nuclear nor are they capable of doing so.[29] There is no doubt that the practical effects of the establishment of the zone are not so significant. France continued to conduct nuclear tests in the region and American armed ships continued their visits to the area even after the treaty came into force. But perhaps what Rarotonga intended was to preserve the *status quo*. In the words of Greg Fry, it was mainly centred on putting obstacles in the way of future proliferation of nuclear weapons rather than changing the existing situation.[30] In another article the same author explains, the idea was to prohibit weapons involvement before it happens and to base the prohibition on the mutual interest of the nuclear-weapon powers in containing the geographical spread of their arms competition.[31]

Another weakness of the Treaty of Rarotonga is that it frees the region of only one aspect of the nuclear arsenals, namely, warheads. This limitation on nuclear weapons legitimises all other kinds of nuclear involvement. For example, prohibition on testing refers only to nuclear warheads, not to missiles. The treaty also does not prevent the hosting of nuclear weapons-related bases in the region. The United States, for example, maintains three bases in Australia.[32] As American analysts William Atkin and Richard Fieldhouse pointed out, "the main flaw in existing nuclear-free policies or nuclear-free-zone proposals is that they set up a system whereby non-nuclear means nothing but the absence of nuclear warheads . . . .[33] Regarding this point, we should always keep in our minds that the makers of South Pacific Nuclear Free Zone Treaty had to adjust to the existing security arrangement in the region; particularly with ANZUS. As Australian foreign minister, Bill Hayden, remarked, the treaty does not seem to undermine the factors that have created and sustained the very favourable security environment which South Pacific enjoys. It does not, in any way, conflict with the Australia's defence arrangements, notably ANZUS.[34] Ramesh Thakur also wrote in an article, before the signing of the treaty, that the most troublesome issue is the insistence that a nuclear-free zone should not disturb existing security arrangement in South Pacific.[35] David Lange has also highlighted this limitation. He observed that "the treaty's limitations are obvious." The most important issue is the existing security arrangement in the region. The treaty could not have disturbed it and as it would have been a major stumbling block.[36]

Yet another weakness of the treaty is that it does not prohibit the transit of nuclear weapons through the region, and the question of port calls has been left open for each country to decide. This could result in stationing also.[37] Regarding this weakness also, David Lange has remarked, "another criterion which demanded respect was the law of the high seas and straits and the Treaty of Rarotongo had to respect that."[38]

Thus, there is no doubt that the Treaty of Rarotonga is a much watered-down version of the nuclear-free zone idea which was initially proposed for the region, but this by no way means that it is meaningless. What is to be kept in mind is that though the Australian proposal took care of its partners and of its own security interest whether it was the question of transit of nuclear-armed ships or the issue of American bases in the region or export of uranium, the treaty does represent a regional response to the issue of arms control and disarmament. It demonstrates the concern of the countries of

South Pacific to protect the fragile ecosystem of the region and the globe. It also reflects the emergence of the South Pacific Forum as a mature, assertive and responsible regional authority. In the words of Jozel Goldbast; "the treaty is a modest but significant contribution to the cause of nuclear non-proliferation."[39] Reflection of this statement can also be seen in the ambitious objectives described in the Preamble to the Treaty.[40] The 5th and the 10th paragraphs reflect the best intention of the authors of the treaty. While the 5th paragraph seeks "to ensure that the bounty and the beauty of the land and the sea in the region shall remain heritage of their people", the 10th paragraph lays down a provision "to keep the region free from environmental pollution". Another point which the Preamble highlights is that the disarmament measures can contribute to global efforts to reverse the nuclear arms race and to promote the national security of adhering states. In the words of Ramesh Thakur, "regional disarmament can at best supplement universal disarmament".[41] The hope is that each step forward towards creating a nuclear-free zone will lessen the suspicion and distrust that underlay the arms race.

Apart from the higher goals of universal disarmament at which the treaty aims at, one immediate event of significance which accompanied the signing of the treaty was the impetus it gave to the regionalism and regional co-operation in South Pacific. It also reflected the assertion of political independence by the Forum Island Countries. They stood against their former colonial masters, on whom they were heavily dependent for aid and assistance. This in itself was a significant development. It is the most effective case of transforming popular sentiment into government policies.

However, the importance of the treaty cannot be over emphasised as has been discussed in Chapter 1. Such efforts should be seen as the beginning of a process leading to the final goal of nuclear disarmament. But since it is a process it should be continued, strengthened and reinforced with the changing times. Since the French nuclear tests have stopped and nuclear powers have signed the treaty, it should move forward for a more stricter regime. It should expand geographically to include American bases like Guam and Palau. Further, a complete ban on the peaceful use of nuclear energy could also be considered, as sacrificing the right to possess peaceful nuclear energy would harm the parties very little while significantly enhancing the validity of the zone. Other than these the transit issue can also be taken up in international forum at least for discussion and consensus building at the initial stage. If UNCLOS also proposes 'peaceful use of high seas' then the ban on transit of nuclear armed ships would not go totally against it. Transit of nuclear armed ships cannot be considered as the 'innocent passage'. A debate on this issue should be opened in different forums. Although the above mentioned goals are difficult to achieve they are important and crucial for the treaties' continuing relevance.

REFERENCES

1. Pande, Savita, "Treaty of Rarotonga: Nuclear Free South Pacific?", *Strategic Analysis*, Vol.21, No.2, May 1998, p.195.
2. Greg Fry, *Working Paper No.75*, ISDS, ANU, Canberrah, pp. 25-27.
3. See Keesing's *Record of World Events*, 1983 (Keynshan, Bristol), Vol.29, No.1, p.32571A.
4. Ibid., p.1986.
5. Paul F. Power, "The South Pacific Nuclear Free Zone", Pacific Affairs, Vol. 59, No.3, Fall 1986, p. 456.
6. Text of the Treaty see Apendix V
7. Carasales, Julio Cesar, "The South Pacific Nuclear Free Zone Rarotonga on the step of Tlatelolco", *World Review*, Vol.26. No.4, December 1987, p.60.
8. Ibid, p.69.
9. See for details, Lini, Walter, "Australia and the South Pacific: A Vanuatu Perspective".
10. Paul F. Power, no.5, p. 465.
11. Thakur, Ramesh, In *Defence of New Zealand: Foreign Policy Choices in the Nuclear Age* (Boulder and London: Westview Press, 1986), p.171.

12. Taken from Mackay, Don J., "Some Legal Aspects of the South Pacific Nuclear Free Zone Treaty", *UN Regional Disarmament Workshop for Asia Pacific* (New York: United Nations, 1991), p.185.
13. Taken from Kaul, Man Mohini, Pearls in the Ocean (New Delhi: UBSP, 1993) p.86.
14. *Washington Post*, 6 February 1987.
15. *Times of India*, 26 March 1996.
16. Power, Paul F., pp.456-66.
17. News Release issued by Minister for Foreign Affairs, Bill Hayden, *AFAR*, Vol.58, No.3, March 1987, p.170.
18. *The Telegraph* (Calcutta), 24 October 1995, p.4.
19. *The News Time*, 29 October 1995.
20. *The Telegraph*, 26 October 1995.
21. *The Guardian*, 4 February 1996.
22. Ibid.
23. *The Telegraph*, 26 October 1995.
24. *International Herald Tribune*, 20 October 1995.
25. *See* Greg Fry, "Towards a South Pacific Nuclear Free Zone", *Bulletin of Atomic Scientists*, Vol.41, No.6, 1985.
26. See Dibb, Paul, "Soviet Strategy towards Australia, New Zealand and South West Pacific", *Australian Outlook*, Vol.29, August 1985.
27. New Release Issued by the Minister of Foreign Affairs Bill Hayden, *AFAR*, Vol.57, No.72, December 1986, pp.1120-21.
28. News Release Issued by Acting Minister for Foreign Affairs Lionel Bowen, *AFAR*, Vol.58, No.2, February 1987, pp.97-98.
29. Mohan, Raja, "The Treaty of Rarotonga", *Strategic Analysis*, Vol. 29, No.7, October 1986, p. 727.
30. Fry, Grey, p.38.
31. Fry, Grey, " The South Pacific Nuclear Free Zone, *SIPRI*, 1986, p. 506.
32. Mohan, Raja, pp.728-29.
33. Arkin William M., and Fieldhouse, Richard W., *Bulletin of Atomic Scientists*, Vol.41, No.6, June/July 1985, p.13.
34. Foreign Minister Hayden, Bill, Speech in Parliament, 5 June 1986.
35. Thakur, Ramesh, "Nuclear Weapon Free South Pacific: A New Zealand Perspective", *Pacific Affairs*, Vol. 58, No.2, Summer 1985, p. 223.
36. Lange, David, *Nuclear Free-The New Zealand Way* (Penguin Books, 1991), p. 170-71.

37. Mohan, Raja, pp.728-29.
38. Lange, pp.170-71.
39. Jozel Goldblast, "The Treaty of Rarotonga: A Modest but Significant Contribution to the Nuclear Non-proliferation", *Transitional Perspective*, Vol.13, No.2, pp.25-27.
40. *See* Text of the Treaty, preamble.
41. Thakur, pp.235-36.

12

Conclusion

As we have discussed in the previous chapters, nuclear-free zones are regional efforts of nuclear non proliferation which can arrest the spread of nuclear weapons as the first step towards eventual elimination of nuclear weapons from the face of the earth. The basic aim of nuclear weapon-free zone is to prevent the introduction of these weapons into areas as yet free from them. It requires non nuclear weapon states of specific region to come together to renounce the introduction, development, deployment or use of nuclear devices in their region. It is an important endeavour keeping in view the breakdown of the bi-polar world and fear of possible increase in the number of states having nuclear weapons and use of these weapons in the regional conflict situations. Moreover, the regional conflicts could also sometimes trigger big power intervention bringing along nuclear danger. The initiative for a nuclear-free zone then becomes a strategy for ruling out conflicts by nuclear means and reducing the threat perception.

The nuclear-free zones ensure that designated zones remain free from big power hegemonic rivalries leading to deployment of nuclear weapons in the region. In this regard, nuclear-free zones also restrict the geographical proliferation of nuclear weapons. Moreover, the establishment of nuclear weapon-free zones encourages 'roll back' proliferation where it has already occurred such as removal of nuclear bases and restricting the movement of nuclear powered ships and aircrafts in the particular region. Setting up such a zone is thus a preventive step aimed at avoiding the neuclearisation of conflicts in the region concerned.

Another important feature of the nuclear-free zones is that they demand respect of nuclear weapon states for the non-nuclear states of the particular region and call on nuclear weapon states to undertake not to use or threaten to use nuclear weapons against them. This negative guarantee given by the nuclear weapon states becomes all the more important in the background of absence of such an assurance in the Non Proliferation Treaty. Although under article VI of the NPT nuclear weapon signatories promised total global disarmament but have failed to abide by the letter and sprit of the clause . In this way, nuclear-free zones fill the gap created by NPT between nuclear weapon states and non nuclear weapon states.

A final positive feature of nuclear-free zones is that the very process of creating such a zone gives a boost to the regionalism and regional cooperation in that region. The negotiating process with its direct contacts and increased communication can help allay suspicions and build confidence among neighbours. Wherever such treaties have come into existence a vigour has been added to the regional organizations and re-integration of individual states into regional subsystems.

However, it is important to see nuclear-free zones as a process rather than an end in itself. Such zones present a potentially effective option to supplement international effort to prevent the proliferation of nuclear weapons.

Nevertheless, to evolve a consensus among all the regional states and eventually to establish a zone free of nuclear weapons is not at all an easy task. Numerous obstacles may impede the establishment and the effective implementation of nuclear-free zones such as, the presence of any nuclear capable or nuclear weapon state in the region. As we have already seen in the case of Latin American nuclear weapon-free zone where the two most important countries of the region-Brazil and Argentina rendered the Treaty of Tlatelolco useless because of their nuclear policies. Also in the case of African nuclear weapon-free zone where the proposal to create such a zone could take shape only when south Africa renounced its nuclear policies. The second obstacle may exist in the form of presence of intra-regional disputes. There could be no better example than South Asia in this regard. To some extent this applies to South East Asia also where the Vietnamese occupation of Cambodia created an atmosphere of acute distrust and suspicion which eventually became a major impediment in the realization of ZOPFAN and SEANWFZ. The third problem which could hinder the effective implementation of nuclear weapon-free zones is their relations with international law. In the case of law of the sea which guarantees freedom of navigation on high seas, it becomes difficult to prohibit the transit of nuclear powered ships in any particular zone. In the South Pacific regional countries initially intended to ban the nuclear powered ships into the zonal area (after the lead taken by New Zealand) but they could not do so. In view of this, it is imperative to redifine the relationship between international laws and nuclear weapon-free zones. Some amendments in the law of the sea *vis-a-vis* the nuclear weapon-free zones would be a welcome step keeping in mind the higher goals of disarmament and nuclear non-proliferation. The fourth problem may arise with regard to the existing security arrangements and the security treaties in the area. It is expected that the nuclear weapon-free zone should respect the existing security arrangements in the region. In the South Pacific the existence of ANZUS security umbrella was the greatest concern for the advocates of nuclear-free zone and ultimately the treaty had to accommodate the security interest of the ANZUS partners. In the case of Southeast Asia this existing power balance made it difficult at first to workout any regional security arrangement and then its effective realization also. Before the proposal forZOPFAN was initiated the regional countries, due to internal, regional, and extra-regional threats and in the absence of adequate resources to protect themselves, had allied with the external powers. As ZOPFAN like security arrangement required a radical change in the existing security pattern, the idea could not take a practical shape. Ironically in the very year of its announcement-1971, Malaysia which was also the initiator of the idea, entered into Five Power Defence Arrangement with Britain (and Australia, New Zealand and Singapore) in continuation with its security relations. Finally, the presence of great powers (particularly the nuclear weapon states), their hegemonic international policies and rivalry also obstruct the formation of nuclear weapon-free zone or the zone of peace. Southeast Asia can be the most suitable example in this regard. The war in Indo-China attracted the intervention by the two super powers and China and ultimately the conflict turned into an extension of the great power rivalry.

In view of the above mentioned problems, the establishment and successful implementation of nuclear weapon-free zones, no doubt, is a difficult task but if all the regional states recognize the need for such a zone in their region, that itself provides the common ground on the basis of which a compromise for the agreement can be reached. In the cases of Southeast Asia and the South Pacific also, the proposals for the establishment of nuclear weapon-free zone have come from within the region and have their origin in the basic survival urge of the countries of both the regions. The proposals emerged out of the desire of these countries to denuclearize and demilitarize their respective regions.

As we have already seen in the case of Southeast Asia, that this region became an arena of superpower rivalry since the start of the cold war only. The region was exposed to the intervention of the external powers to such an extent that the regional conflict almost became an extension of the conflict between the two super powers. The US policies towards Southeast Asia were mainly guided by the so called 'communist threat'. Hence the Vietnam war and the Cambodian crisis pulled Washington more and more into Southeast Asia. Other than the communist dominated countries the US presence was also viewed vital by the regional countries for the security and stability. On the other hand the Soviet presence too was mainly dominated by its policies towards the other bloc which led to its heavy involvement in the Indo-China conflict. Another major external power which was active in Southeast Asia was China. China was closely involved in the Cambodian conflict and at present, it is also considered as one of the parties to the conflicting claims over the South China Sea.

The situation arising from the intensified conflict in Indo-China due to excessive external intervention also polarized the regional countries and completely divided the region along two blocs. But at the same time the situation intensified the need for a regional security order. The formation of ASEAN itself gave a boost to the regional security order also. The major initiative in this direction came as ASEAN's proposal for ZOPFAN. Since the nuclear threat was implicit in this situation the proposal for a nuclear-free zone developed as a component of ZOPFAN. However, the later development such as the communist invasion of Cambodia in 1978 and the Sino-Vietnamese antagonism gave a major blow to ZOPFAN and the ASEAN's agenda for a regional security arrangement. Even after the end of cold war and the settlement of Indo-China conflict the ZOPFAN still has failed to take a practical shape due to the South China Sea disputes. However, nuclear weapon-free zone treaty has been adopted as an interim arrangement. The establishment of Southeast Asia Nuclear Weapon-free Zone, as an essential component of ZOPFAN will contribute towards strengthening the security of states within the zone and towards enhancing international peace and security as a whole.

As far as the South Pacific is concerned, unlike Southeast Asia this region faced a direct nuclear threat due to nuclear testing and dumping of nuclear wastes. However, there was an atmosphere of relative peace and harmony in the region and any conflict or threat of such conflict among the regional states was absent. Moreover, there has always been a fair degree cooperation and mutual understanding among them. Neverthless, security threats do exit, but has a non-conventional dimension such as environment, economic vulnerability, etc. The nuclear threat also was mainly related to the environment and human security. The treaty of Rarotonga was a result of a strong anti nuclear sentiment since the beginning of the nuclear testing by the US. In spite of the fact that almost all the island countries were under developed and heavily depended on aid and assistance and there was a great deal of opposition from the nuclear weapon states, all the regional countries stood united on the nuclear issue and the anti nuclear movement culminated into the signing of the South Pacific Nuclear Free Zone Treaty.

Thus, it may be stated that the motives behind the call for nuclear weapon-free zones vary from region to region but the factor common to all zones is the belief among the states that they would be more secure, if their regions are free of nuclear weapons. The concepts of nuclear weapon-free zones have continuously evolved with the establishment of each zone and should be viewed as a process in the direction of achieving the goal of global peace and disarmament.

Kulala Lumpur Declaration (ZOPEAN Declaration, 1971)

We, the Foreign Ministers of

INDONESIA,

MALAYSIA,

THE PHILIPPINES,

SINGAPORE

and

the Special Envoy of the National Executive Council of THAILAND

FIRMLY BELIEVING in the merits of regional cooperation which has drawn our countries to cooperate together in the economic, social and cultural fields in the Association of Southeast Asian Nations;

DESIROUS of bringing about a relaxation of international tension and of achieving a lasting peace in Southeast Asia;

INSPIRED by the worthy aims and objectives of the United Nations, in particular by the principles of respect for the sovereignty and territorial integrity of all States, abstention from the threat or use of force, peaceful settlement of international disputes, equal rights and self-determination and non-interference in the internal affairs of States;

BELIEVING in the continuing validity of the 'Declaration on the Promotion of World Peace and Cooperation' of the Bandung

Conference of 1955, which, among others, enunciates the principles by which States may coexist peacefully;

RECOGNISING the right of every state, large or small, to lead its national existence free from outside interference in its internal affairs as this interference will adversely affects its freedom, independence and integrity;

DEDICATED to the maintenance of peace, freedom and independence unimpaired;

BELIEVING in the need to meet present challenges and new developments by cooperating with all peace and freedom loving nations, both within and outside the region, in the furtherance of world peace, stability and harmony;

COGNIZANT of the significant trend towards establishing nuclear-free zones, as in the 'Treaty for the Prohibition of Nuclear Weapons in Latin America' and the Lusaka Declaration proclaiming Africa a nuclear-free zone, for the purpose of promoting world peace and security by reducing the areas of international conflicts and tensions;

REITERATING our commitment to the principle in the Bangkok Declaration which established ASEAN in 1967, that the countries of Southeast Asia share a primary responsibility for strengthening the economic and social stability of the region and ensuring their peaceful and progressive national development, and that they are determined to ensure their stability and security from external interference in any form or manifestation in order to preserve their national identities in accordance with the ideals and aspirations of their peoples;

AGREEING that the neutralization of Southeast Asia is a desirable objective and that we should explore ways and means of bringing about its realisation, and

CONVINCED that the time is propitious for joint action to give effective expression to the deeply felt desire of the peoples of Southeast Asia to ensure the conditions of peace and stability indispensable to their independence and their economic and social well-being;

DO HEREBY STATE:

1. that Indonesia, Malaysia, the Philippines, Singapore and Thailand are determined to exert initially necessary efforts to secure the recognition of, and respect for, Southeast Asia as a Zone of Peace, Freedom and Neutrality, free from any form or manner of interference by outside powers;
2. that Southeast Asian countries should make concerted efforts to broaden the areas of cooperation which would contribute to their strength, solidarity and closer relationship.

Done in Kuala Lumpur on Saturday, the 27th of November 1971.

On behalf of the Republic of Indonesia

(Adam Malik)
Minister of Foreign Affairs.

On behalf of Malasia

(Tun Abdul Razak bin Dato Hussein)
Prime Minister and Minister of Foreign Affairs.

On behalf of the Republic of the Philippines

(Carlos P. Romulo)
Secretary of Foreign Affairs.

Treaty of Amity and Cooperation in Southeast Asia

Denpasar, Bali, 24, February 1976

PREAMBLE

The High Contracting Parties

Conscious of the existing of history, geography and culture, which have bound their peoples together;

Anxious to promote regional peace and stability through abiding respect for justice and the rule of law and enhancing resilience in their relations;

Desiring to enhance peace, friendship and mutual cooperation on matters affecting Southeast Asia consistent with the spirit and principles of the Charter of the United Nations, the Ten principles adopted by the Asian-African Conference in Bandung on 25 April 1955, the Declaration of the Association of Southeast Asian Nations signed in Bangkok on 8 August 1967, and the Declaration signed in Kuala Lumpur on 27 November 1971.

Convinced that the settlement of difference or disputes between their countries should be regulated by rational effective and sufficiently flexible procedures, avoiding negative attitudes which might endanger or hinder cooperation.

Believing in the need for cooperation with all peace-loving nations, both within and outside Southeast Asia, in the furtherance of world peace, stability and harmony.

Solemnly agree to enter into a Treaty of Amity and Cooperation as follows:

CHAPTER I: PURPOSE AND PRINCIPLES

Article I

The purpose of this Treaty is to promote perpetual peace, everlasting amity and cooperation among their peoples which would contribute to their strength, solidarity and closer relationship.

Article II

In their relations with one another, the High Contracting Parties shall be guided by the following fundamental principles:

a) Mutual respect for the independence, sovereignty, equality, territorial integrity and national identify of all nations;

b) The right of every State to lead its national existence free from external interference, subversion or coersion;

c) Non-interference in the internal affairs of one another;

d) Settlement of differences or disputes by peaceful means;

e) Renunciation of the threat or use of force;

f) Effective cooperation among themselves.

CHAPTER II: AMITY

Article III

In pursuance of the purpose of this Treaty, the High Contracting Parties shall endeavour to develop and strengthen the traditional, cultural and historical ties of friendship, good neighbourliness and cooperation which bind them together and shall fulfil in good faith the obligations assumed under this Treaty. In order to promote closer understanding among them, the High Contracting Parties shall encourage and facilitate contract and intercourse among their peoples.

CHAPTER III: COOPERATION

Article IV

The High Contracting Parties shall promote active cooperation in the economic, social, technical, scientific and administrative fields as well as in matters of common ideals and aspiration of international peace and stability in the region and all other matters of common interest.

Article V

Pursuant to Article 4, the High Contracting Parties shall exert their maximum efforts multilaterally as well as bilaterally on the basis of equality, non-discrimination and mutual benefit.

Article VI

The High Contracting Parties shall collaborate for the acceleration of the economic growth in the region in order to strengthen the foundation for a prosperous and peaceful community of nations in Southeast Asia. To this end, they shall promote the greater utilisation of their agriculture and industries, the expansion of their trade and the improvement of their economic infrastructure for the mutual benefit of their peoples. In this regard, they shall continue to explore all avenues for close and beneficial cooperation with other States as well as international and regional organisations outside the region.

Article VIII

The High Contracting Parties, in order to achieve social justice and to raise the standards of living of the peoples of the region, shall intensify economic cooperation. For this purpose, they shall adopt appropriate regional strategies for economic development and mutual assistance.

Article VIII

The High Contracting Parties shall strive to achieve the closet cooperation on the widest scale and shall seek to provide assistance to one another in the form training and research facilities in the social, cultural, technical, scientific and administrative fields.

Article IX

The High Contracting Parties shall endeavour to foster cooperation in the furtherance of the cause of peace, harmony and stability in the region. To this end, the High Contracting Parties shall maintain regular contacts and consultations with one another on international and regional matters with a view to coordinating their views, actions and policies.

Article X

Each High Contracting Party shall not in any manner or form participate in any activity which shall constitute a threat to the political and economic stability, sovereignty, or territorial integrity of another High Contracting Party.

Article XI

The High Contracting Parties shall endeavour to strengthen their respective national resilience in their politicl, economic, socio-cultural as well as security fields in conformity with their respective ideals and aspirations, free from external interference as well as internal subversive activities in order to preserve their respective national identities.

Article XII

The High Contracting Parties in their efforts to achieve regional prosperity and security, shall endeavour to cooperate in all fields for the promotion of regional resilience based on the principles of self-confidence, self-reliance, mutual respect, cooperation and solidarity which will constitute the foundation for a strong and viable community of nations in Southeast Asia.

CHAPTER IV: PACIFIC SETTLEMENT OF DISPUTES

Article XIII

The High Contracting Parties shall have the determination and good faith to prevent disputes from arising. In case disputes on matters directly affecting them should arise, especially disputes likely to disturb regional peace and harmony, they shall refrain from the threat or use of force and shall at all times settle such disputes among themselves through friendly negotiations.

Article XIV

To settle disputes through regional processes, the High Contracting Party shall constitute, as a continuing body, a High Council Comprising of Representatives at the Ministerial level from each of the High Contracting Parties to take cognizance of the existence of disputes or situation likely to disturb regional peace and harmony.

Article XV

In the event no solution is reached through direct negotiations, the High Council shall take cognizance of the dispute or the situation and shall recommend to the parties in dispute appropriate means of settlement, such as good offices, mediation, inquiry of conciliation. The High Council may however offer its good offices, or upon agreement of the parties in dispute, constitute itself into a committee of mediation, inquiry or conciliation. When deemed necessary, the High Council shall recommend appropriate measures for the prevention of a deterioration of the dispute or the situation.

Article XVI

The foregoing provision of this Chapter shall not apply to dispute unless all the parties to the dispute agree to their application to that dispute. However, this shall not preclude the other High Contracting Parties not party to the dispute from offering all possible assistance to settle the said dispute. Parties to the dispute should be well disposed towards such offers of assistance.

Article XVII

Nothing in this Treaty shall preclude recourse to the modes of peaceful settlement contained in Article 33 (1) of the Charter of the United Nations. The High Contracting Parties which are parties to dispute should be encouraged to take initiatives to solve it by friendly negotiations before resorting to the other procedures provided for in the Charter of the United Nations.

CHAPTER V: GENERAL PROVISIONS

Article XVIII

This Treaty shall be signed by the Republic of Indonesia, Malaysia, the Republic of the Philippines, the Republic of

Singapore and the Kingdom of Thailand. It shall be ratified in accordance with the constitutional procedures of each signatory State.

It shall be open for accession by other States in Southeast Asia.

Article XIX

This Treaty shall enter into force on the date of the deposit of the fifth instrument of ratification with the Governments of the signatory States which are designated Depositories of this Treaty and of the instruments of ratification or accession.

Article XX

This Treaty is drawn up in the official languages of the High Contracting Parties, all of which are equally authoritative. There shall be an agreed common translation of the texts in the English language. Any divergent interpretation of the common text shall be settled by negotiation.

In faith thereof the High Contracting Parties have signed the Treaty and have hereto affixed their seals.

Done at Denpasar, Bali, this twenty-fourth day of February in the year one thousand nine hundred and seventy-six.

For Republic of Indonesia:

Sd. Soeharto
President

For Malaysia

Sd. Datuk Hussein Onn
Prime Minister

For the Republic of the Philippines

Sd. Ferdinand E. Marcos
President

For the Republic of Singapore

Sd. Lee Kuan Yew
Prime Minister

For the Kingdom of Thailand

Sd. Kukrit Pramoj
Prime Minister

Declaration of ASEAN Concord

A Common Bond Existing among the Member States of the Association of Southeast Asian Nations

The President of the Republic of Indonesia, the Prime Minister of Malaysia, the President of the Republic of the Philippines, the Prime Minister of the Republic of Singapore and the Prime Minister of the Kingdom of Thailand

REAFFIRM their commitment to the Declarations of Bandung, Bangkok and Kuala Lumpur, and the Charter of the United Nations;

ENDEAVOUR to promote peace, progress, prosperity and the welfare of the peoples of member states;

UNDERTAKE to consolidate the achievement of ASEAN and expand ASEAN cooperation in the economic, social, cultural and political fields;

DO HEREBY DECLARE:

ASEAN Cooperation shall take into account, among others, the following objectives and principles in the pursuit of political stability:

1. The stability of each member state and of the ASEAN region is an essential contribution to international peace and security. Each member state resolves to eliminate threats posed by subversion to its stability, thus strengthening national and ASEAN resilience.

2. Member states, individually and collectively, shall take active steps for the early establishment of the Zone of Peace, Freedom and Neutrality.

Treaty on the Southeast Asia Nuclear Weapon-free Zone (Bangkok Treaty)

entered in force 1995

The States Parties to this Treaty:

DESIRING to contribute to the realisation of the purposes and principles of the Charter of the United Nations;

DETERMINED to take concrete action which will contribute to the progress towards general and complete disarmament of nuclear weapons, and to the promotion of international peace and security;

REAFFIRMING the desire of the Southeast Asian States to maintain peace and stability in the region in the spirit of peaceful coexistence and mutual understanding and cooperation as enunciated in various communiques, declarations and other legal instruments;

RECALLING the Declaration on the Zone of Peace, Freedom and Neutrality (ZOPFAN) signed in Kuala Lumpur on 27 November 1971 and the Programme of Action on ZOPFAN adopted at the 26th ASEAN Ministerial Meeting in Singapore in July 1993;

CONVINCED that the establishment of a Southeast Asia Nuclear Weapon-Free Zone, as an essential component of the ZOPPAN, will contribute towards strengthening the security of

States within the Zone and towards enhancing international peace and security as a whole;

REAFFIRMING the importance of the Treaty on the Non-Proliferation of Nuclear Weapons (NPT) in preventing the proliferation of nuclear weapons and in contributing towards international peace and security;

RECALLING Article VII of the NPT which recognises the right of any group of States to conclude regional treaties in order to assume the total absence of nuclear weapons in their respective territories;

RECALLING the Final Document of the Tenth Special Session of the United Nations General Assembly which encourages the establishment of nuclear weapon-free zones;

RECALLING the Principles and Objectives for Nuclear Non-Proliferation and Disarmament, adopted at the 1995 Review and Extension Conference of the Parties to the NPT, that the cooperation of all the nuclear-weapon States and their respect and support for the relevant protocols is important for the maximum effectiveness of this nuclear weapon-free zone treaty and its relevant protocols.

DETERMINED to protect the region from environmental pollution and the hazards posed by radioactive wastes and other radioactive material;

HAVE AGREED as follows:

ARTICLE 1

Use of Forms

For the purposes of this Treaty and its Protocol:

a) "Southeast Asia Nuclear Weapon-Free Zone", hereinafter referred to as the "Zone", means the area comprising the territories of all States in Southeast Asia, namely, Brunei

Darussalam, Cambodia, Indonesia, Laos, Malaysia, Myanmar, the Philippines, Singapore, Thailand and Vietnam, and their respective continental shelves and Exclusive Economic Zones (EEZ);

b) "territory" means the land territory, internal waters, territorial sea, archipelagic waters, the seabed and the sub-soil thereof and the airspace above them;

c) "nuclear weapon" means any explosive device capable of releasing nuclear energy in an uncontrolled manner but does not include the means of transport or delivery of such device if separable from and not an indivisible part thereof;

d) "station" means to deploy, emplace, implant, install, stockpile or store;

e) "radioactive material" means material that contains radionuclides above clearance or exemption levels recommended by the International Atomic Energy Agency (IAEA);

f) "radioactive wastes" means material that contains or is contaminated with radionuclides at concentrations or activities greater than clearance levels recommended by the IAEA and for which no use is foreseen; and

g) "dumping" means

   i) any deliberate disposal at sea, including seabed and subsoil insertion, of radioactive wastes or other matter from vessels, aircraft, platforms or other man-made structures at sea, and

   ii) any deliberate disposal at sea, including seabed and subsoil insertion, of vessels, aircraft, platforms or other man-made structures at sea, containing radioactive material, but does not include the disposal of wastes or other matter incidental to, or derived from the normal operations of vessels, aircraft, platforms or other man-made structures at sea and their equipment, other than wastes or other matter transported by or to vessels, aircraft, platforms or other man-made structures at sea, operating for the purpose of disposal of such matter or derived from the treatment of such wastes or other matter on such vessels, aircraft, platforms or structures.

ARTICLE 2

Application of Treaty

1. This Treaty and its Protocol shall apply to the territories, continental selves, and EEZ of the States Parties within the Zone in which the Treaty is in force.
2. Nothing in this Treaty shall prejudice the rights or the exercise of these rights by any State under the provisions of the United Nations Convention on the Law of the Sea of 1982, in particular with regard to freedom of the high seas, rights of innocent passage, archipelagic sea lanes passage or transit passage of ships and aircraft, and consistent with the Charter of the United Nations.

ARTICLE 3

Basic Understanding

1. Each State Party undertakes not to, anywhere inside or outside the Zone:

   a) develop, manufacture or otherwise acquire, possess or have control over nuclear weapons;
   b) station or transport nuclear weapons by any means; or
   c) test or use nuclear weapons.

2. Each State Party also undertakes not to allow, in its territory, any other State to:
   a) develop, manufacture or otherwise acquire, possess or have control over nuclear weapons;
   b) station nuclear weapons; or
   c) test or use nuclear weapons.

3. Each State Party also undertakes not to:
   a) dump at sea or discharge into the atmosphere anywhere within the Zone any radioactive material or wastes;
   b) dispose radioactive material or wastes on land in the territory of or under the jurisdiction of other States except as stipulated in Paragraph 2 (e) of Article 4; or
   c) allow, within its territory, any other State to dump at sea or discharge into the atmosphere any radioactive material or wastes.

4. Each State Party undertakes not to:
   a) seek or receive any assistance in the Commission of any act in violation of the provisions of Paragraphs 1, 2 and 3 of this Article; or
   b) take any action to assist or encourage the Commission of any act in violation of the provisions of Paragraphs 1, 2 and 3 of this Article.

ARTICLE 4

Use of Nuclear Energy for Peaceful Purposes

1. Nothing in this Treaty shall prejudice the right of the States Parties to use nuclear energy, in particular for their economic development and social progress.

2. Each State Party therefore undertakes:

   a) to use exclusively for peaceful purposes nuclear material and facilities which are within its territory and areas under its jurisdiction and control;

   b) prior to embarking on its peaceful nuclear energy programme, to subject its programme to rigorous nuclear safety assessment conforming to guidelines and standards recommended by the IAEA for the protection of health and minimisation of danger to life and property in accordance with Paragraph 6 of Article III of the Statute of the IAEA;

   c) upon request, to make available to another State Party the assessment except information relating to personal data, information protected by intellectual property rights or by industrial or commercial confidentiality, and information relating to national security;

   d) to support the continued effectiveness of the international non-proliferation system based on the Treaty on the Non-Proliferation of Nuclear Weapons (NPT) and the IAEA safeguard system; and

   e) to dispose radioactive wastes and other radioactive material in accordance with IAEA standards and procedures on land within its territory or on land within the territory of another State which has consented to such disposal.

3. Each State Party further undertakes not to provide source or special fissionable material, or equipment or material especially designed or prepared for the processing, use or production of special fissionable material to:

   a) any non-nuclear weapon State except under conditions subject to the safeguards required by Paragraph 1 of Article III of the NPT; or

b) any nuclear weapon State except in conformity with applicable safeguards agreements with the IAEA.

ARTICLE 5

IAEA Safeguards

Each State Party which has not done so shall conclude an agreement with the IAEA for the application of full scope safeguards to its peaceful nuclear activities not later than eighteen months after the entry into force for that State Party of the Treaty.

ARTICLE 6

Early Notification of a Nuclear Accident

Each State Party which has not acceded to the Convention on Early Notification of a Nuclear Accident shall endeavour to do so.

ARTICLE 7

Foreign Ships and Aircraft

Each State Party, on being notified, may decide for itself whether to allow visits by foreign ships and aircraft to its ports and airfields, transit of its airspace by foreign aircraft, and navigation by foreign ships through its territorial sea or archipelagic waters and overflight of foreign aircraft above those waters in a manner not governed by the rights of innocent passage, archipelagic sea lanes passage or transit passage.

ARTICLE 8

Establishment of the Commission for the Southeast Asia Nuclear Weapon-free Zone

1. There is hereby established a Commission for the Southeast Asia Nuclear Weapon-Free Zone, hereinafter referred to as the "Commission".
2. All States Parties are *ipso facto* members or the Commission. Each State Party shall be represented by its Foreign Minister or his representative accompanied by alternates and advisers.
3. The function of the Commission shall be to oversee the implementation of this Treaty and ensure compliance with its provisions.
4. The Commission shall meet as and when necessary in accordance with the provisions of this Treaty including upon the request of any State Party. As far as possible, the Commission shall meet in conjunction with the ASEAN Ministerial Meeting.
5. At the beginning of each meeting, the Commission shall elect its Chairman and such other officers as may be required. They shall hold office until a new Chairman and other officers are elected at the next meeting.
6. Unless otherwise provided for in this Treaty, two-thirds of the members of the Commission shall be present to constitute a quorum.
7. Each member of the Commission shall have one vote.
8. Except as provided for in this Treaty, decisions of the Commission shall be taken by consensus or, failing consensus, by a two-thirds majority of the members present and voting.
9. The Commission shall, by consensus, agree upon and adopt rules of procedure for itself as well as financial

rules governing its funding and that of its subsidiary organs.

ARTICLE 9

The Executive Committee

1. There is hereby established, as a subsidiary organ of the Commission, the Executive Committee.
2. The Executive Committee shall be composed of all States Parties to this Treaty. Each State Party shall be represented by one senior official as its representative, who may be accompanied by alternates and advisers.
3. The functions of the Executive Committee shall be to:

    a) ensure the proper operation of verification measures in accordance with the provisions on the control system as stipulated in Article 10;
    b) consider and decide on requests for clarification and for a fact-finding mission;
    c) set up a fact-finding mission in accordance with the Annex of this Treaty;
    d) consider and decide on the findings of a fact-finding mission and report to the Commission;
    e) request the Commission to convene a meeting when appropriate and necessary;
    f) conclude such agreements with the IAEA or other international organisations as referred to in Article 18 on behalf of the Commission after being duly authorised to do so by the Commission; and
    g) carry out such other tasks as may, from time to time, be assigned by the Commission.

4. The Executive Committee shall meet as and when necessary for the efficient exercise of its functions. As far as possible, the Executive Committee shall meet in conjunction with the ASEAN Senior Officials Meeting.
5. The Chairman of the Executive Committee shall be the representative of the Chairman of the Commission. Any submission or communication made by a State Party to the Chairman of the Executive Committee shall be disseminated to the other members of the Executive Committee.
6. Two-thirds of the members of the Executive Committee shall be present to constitute a quorum.
7. Each member of the Executive Committee shall have one vote.
8. Decisions of the Executive Committee shall be taken by consensus or, failing consensus, by a two-thirds majority of the members present and voting.

ARTICLE 10

Control System

1. There is hereby established a control system for the purpose of verifying compliance with the obligations of the States Parties under this Treaty.
2. The Control System shall comprise:

   a) the IAEA safeguards system as provided for in Article 5;
   b) report and exchange of information as provided for in Article 11;
   c) request for clarification as provided for in Article 12; and

d) for a fact-finding mission as provided for in Article

ARTICLE 11

Reprot and Exchange of Information

1. Each State Party shall submit reports to the Executive Committee on any significant event within its territory and areas under its jurisdiction and control affecting the implementation of this Treaty.
2. The States Parties may exchange information on matters arising under or in relation to this Treaty.

ARTICLE 12

Request for Clarification

1. Each State Party shall have the right to request another State Party for clarification concerning any situation which may be considered ambiguous or which may give rise to doubts about the compliance of that State Party with this Treaty. It shall inform the Executive Committee of such a request. The requested State Party shall duly respond by providing without delay the necessary information and inform the Executive Committee of its reply to the requesting State Party.
2. Each State Party shall have the right to request the Executive Committee to seek clarification for another State Party concerning any situation which may be considered ambiguous or which may give rise to doubts about compliance of that State Party with this Treaty. Upon receipt of such a request, the Executive Committee shall consult the State Party from which clarification is sought for the purpose of obtaining the clarification requested.

ARTICLE 13

Request for a Fact-finding Mission

A State Party shall have the right to request the Executive Committee to send a fact-finding mission to another State Party in order to clarify and resolve a situation which may be considered ambiguous or which may give rise to doubts about compliance with the provisions of this Treaty, in accordance with the procedure contained in the Annex to this Treaty.

ARTICLE 14

Remedial Measures

1. In case the Executive Committee decide in accordance with the Annex that there is a breach of this Treaty by a State Party, that State Party shall, within a reasonable time, take all steps necessary to bring itself in full compliance with this Treaty and shall promptly inform the Executive Committee of the action taken or proposed to be taken by it.
2. Where a State Party fails or refuses to comply with the provisions of Paragraph 1 of this Article, the Executive Committee shall request the Commission to convene a meeting in accordance with the provisions of Paragraph 3(e) of Article 9.
3. At the meeting convened pursuant to Paragraph 2 of this Article, the Commission shall consider the emergent situation and shall decide on any measure it deems appropriate to cope with the situation, including the submission of the matter to the IAEA and, where the situation might endanger international peace and security, the Security Council and the General Assembly of the United Nations.

4. In the event of breach of the Protocol attached to this Treaty by a State Party to the Protocol, the Executive Committee shall convene a special meeting of the Commission to decide on appropriate measures to be taken.

ARTICLE 15

Signature, Ratification, Accession, Deposit and Registration

1. This Treaty shall be open for signature by all States in Southeast Asia, namely, Brunei Darussalam, Cambodia, Indonesia, Laos, Malaysia, Myanmar, the Philippines, Singapore, Thailand and Vietnam.
2. This Treaty shall be subject to ratification in accordance with the constitutional procedure of the signatory states. The instruments of ratification shall be deposited with the Government of the Kingdom of Thailand which is hereby designated as the Depositary State.
3. This Treaty shall be open for accession. The instruments of accession shall be deposited with the Depositary State.
4. The Depositary State shall inform the other States Parties to this Treaty on the deposit of instruments of ratification or accession.
5. The Depositary State shall register this Treaty and its Protocol pursuant to Article 102 of the Charter of the United Nations.

ARTICLE 16

Entry into Force

1. This Treaty shall enter into force on the date of the deposit of the seventh instrument of ratification and/or accession.

2. For States which ratify or accede to this Treaty after the date of the seventh instrument of ratification or accession, the Treaty shall enter into force on the date of deposit of its instrument of ratification or accession.

ARTICLE 17

Reservations

This Treaty shall not be subject to reservations.

ARTICLE 18

Relations with Other International Organisations

The Commission may conclude such agreements with the IAEA or other international organisations as it considers likely to facilitate the efficient operation of the control system established by this Treaty.

ARTICLE 19

Amendments

1. Any State Party may propose amendments to this Treaty and its Protocol and shall submit its proposals to the Executive Committee, which shall transmit them to all the other States Parties. The Executive Committee shall immediately request the Commission to convene a meeting to examine the proposed amendments. The quorum required for such a meeting shall be all the members of the Commission. Any amendment shall be adopted by a consensus decision of the Commission.
2. Amendments adopted shall enter into force 30 days after the receipt by the Depositary State of the seventh instrument of acceptance from the States Parties.

ARTICLE 20

Review

Ten years after this Treaty enters into force, a meeting of the Commission shall be convened for the purpose of reviewing the operation of the Treaty. A meeting of the Commission for the same purpose may also be convened at anytime thereafter if there is consensus among all its members.

ARTICLE 21

Settlement of Disputes

Any dispute arising from the interpretation of the provisions of this Treaty shall be settled by peaceful means as may be agreed upon by the States Parties to the dispute. If within one month, the parties to the dispute are unable to achieve a peaceful settlement of the dispute by negotiation, mediation, enquiry or conciliation, any of the parties concerned shall, with the prior consent of the other parties concerned, refer the dispute to arbitration or to the International Court of Justice.

ARTICLE 22

Duration and Withdrawal

1. This Treaty shall remain in force indefinitely.
2. In the event of a breach by any State Party of this Treaty essential to the achievement of the objectives of the Treaty, every other State Party shall have the right to withdraw from the Treaty.
3. Withdrawal under Paragraph 2 of Article 22, shall be effected by giving notice twelve months in advance to the members of the Commission.

IN WITNESS WHEREOF, the undersigned have signed this Treaty.

DONE at Bangkok, this fifteenth day of December, one thousand nine hundred and ninety-five, in one original in the English language.

For Brunei Darussalam

Signed: Haji Hassanal Bokiah
Sultan of Brunei Darussalam

For the Kingdom of Cambodia

Signed: Samdech Krom Preah Nordom Ranariddh
Samdech Hun Sen
First Prime Minister
Second Prime Minister

For the Republic of Indonesia

Signed: Soeharto
President

For the Lao People's Democratic Republic

Signed: Khamtay Siphondone
Prime Minister

For Malaysia

Signed: Dr. Mahathir bin Mohamad
Prime Minister

For the Union of Myanmar

Signed: Senior General Thanshwe
Chairman of the State Law and Order
Restoration Council and Prime Minister

For the republic of Philippines

Signed: Fidsel. V. Ramos
President

For the republication of Singapore

Signed: Goh Chok Tong
Prime Minister

For the Kingdom of Thailand

Signed: Banharn Silpa-Archa
Prime Minister

For the Socialsit Republic of Vietnam

Signed: Vo van Kiet
Prime Minister

PROTOCOL TO THE TREATY ON SOUTHEAST ASIA NUCLEAR WEAPON -FREE ZONE

The States- Parties to this Protocol,

Desiring to contribute to efforts towards achieving general and complete disarmament of nuclear weapons, and thereby ensuring international peace and security, including in Southeast Asia;

Noting the Treaty on the Southeast Asia Nuclear Weapon - Free Zone, signed at Bangkok, on the fifteenth day of December, one thousand nine hundred and ninety-five;

Have agreed as follows:

Article 1

Each State Party undertakes to respect the Treaty on the Southeast Asia Nuclear Weapon-Free Zone, hereinafter referred to as the Treaty, and not to contribute to any act which constitutes a violation of the Treaty or its Protocol by State Parties to them.

Article 2

Each State Party undertakes, not to use or threaten to use nuclear weapons against any State Party to the Treaty. It further against any State Party to the Treaty. It further undertakes not to use or threaten to use nuclear weapons within the Southeast Asia Nuclear Weapon Free Zone.

Article 3

This Protocol shall be open for signature by the people's Republic of China, the French Republic, the Russian Federation, the United Kingdom of Great Britain and Nothern Ireland and the United States of America.

Article 4

Each State Party undertakes, by written notification to the Depositary State, to indicate its acceptance or otherwise of any alteration to its obligations under this Protocol that may be brought about by the entry into force of any alteration to its obligations under this Protocol that may be brought about by the entry into force of an amendment to the Treaty pursuant to Article 19 thereof.

Article 5

This Protocol is of a permanent nature and shall remain in force indefinitely, provided that each State Party shall, in exercising its national sovereignty, have the right to withdraw from this Protocol if it decides that extraordinary events, related to the subject-matter of this Protocol, have jeopardised its supreme national interests. It shall give notice of such withdrawal to the Depositary State twelve months in advance. Such notice shall include a statement of the extraordinary events its regards as having jeopardised its supreme national interests.

Article 6

This Protocol shall be subject to ratification.

ARTICLE 7

This Protocol shall enter into force for each State Party on the date of its deposit of its instrument of ratification with the Depositary State. The Depositary State shall inform the other States Parties to the Treaty and to this Protocol on the deposit of instruments of ratification.

In witness whereof the undersigned, being duly authorized by their Governments, have signed this Protocol.

Done at Bangkok this fifteenth day of December, one thousand nine hundred and ninety-five, in one original in the English language.

South Pacific Nuclear-free Zone Treaty

(Rarotonga Treaty)

On 6 August 1985 the South Pacific Forum, a body comprising the independent and self-governing countries of the South Pacific (Australia, the Cook Islands, Fiji, Kiribati, Nauru, New Zealand, Niue, Papua New Guinea, the Solomon Islands, Tonga, Tuvalu, Vanuatu and Western Samoa), endorsed the text of the South Pacific Nuclear Free Zone Treaty and opened it for signature.

The following text of the Treaty with its Annexes was formally communicated to the Director General by the Director of the South Pacific Bureau for Economic Cooperation (SPEC), who has been nominated as Depositary of the Treaty, and is herewith being circulated to all Member States for their information pursuant to a request made by the Director of SPEC.

PREAMBLE

The Parties to this Treaty,

United in their commitment to a world at peace;

Gravely concerned that the continuing nuclear arms race presents the risk of nuclear war which would have devastating consequences for all people;

Convinced that all countries have an obligation to make every effort to achieve the gcal of eliminating nuclear weapons, the terror which they hold for humankind and the threat which they pose to life on earth;

Believing that regional arms control measures can contribute to global efforts to reverse the nuclear arms race and promote the national security of each country in the region and the common security of all;

Determined to ensure, so far as lies within their power, that the bounty and beauty of the land and sea in their region shall remain the heritage of their peoples and their descendants in perpetuity to be enjoyed by all in peace;

Reaffirming the importance of the Treaty on the Non-Proliferation of Nuclear Weapons (NPT) in preventing the proliferation of nuclear weapons and in contributing to world security;

Noting, in particular, that Article VII of the NPT recognizes the right of any group of States to conclude regional treaties in order to assure the total absence of nuclear weapons in their respective territories;

Noting that the prohibitions of emplantation and emplacement of nuclear weapons on the seabed and the ocean floor and in the subsoil thereof contained in the Treaty on the Prohibition of the Emplacement of Nuclear Weapons and Other Weapons of Mass Destruction on the Seabed and the Ocean Floor and in the Subsoil Thereof apply in the South Pacific;

Noting also that the prohibition of testing of nuclear weapons in the atmosphere or under water, including territorial waters or high seas, contained in the Treaty Banning Nuclear Weapon Tests in the Atmosphere, in Outer Space and Under Water applies in the South Pacific;

Determined to keep the region free of environmental pollution by radioactive wastes and other radioactive matter;

Guided by the decision of the Fifteenth South Pacific Forum at Tuvalu that a nuclear free zone should be established in the region at the earliest possible opportunity in accordance with the principles set out in the communique of that meeting;

Have agreed as follows:

ARTICLE 1

Usage of Terms

For the purposes of this Treaty and its Protocols:

a) "South Pacific Nuclear Free Zone" means the areas described in Annex 1 as illustrated by the map attached to that Annex;

b) "territory" means internal waters, territorial sea and archipelagic waters, the seabed and subsoil beneath, the land territory and the airspace above them;

c) "nuclear explosive device" means any nuclear weapon or other explosive device capable of releasing nuclear energy, irrespective of the purpose for which it could be used. The term includes such a weapon or device in unassembled and partly assembled forms, but does not include the means of transport or delivery of such a weapon or device if separable from and not an indivisible part of it;

d) "stationing" means emplantation, emplacement, transportation on land or inland waters, stockpiling, storage, installation and deployment.

ARTICLE 2

Application of the Treaty

1. Except where otherwise specified, this Treaty and its

Protocols shall apply to territory within the South Pacific Nuclear Free Zone.

2. Nothing in this Treaty shall prejudice or in any way affect the rights, or the exercise of the rights, of any State under international law with regard to freedom of the seas.

ARTICLE 3

Renunciation of Nuclear Explosive Devices

Each Party undertakes:

a) not to manufacture or otherwise acquire, possess or have control over any nuclear explosive device by any means anywhere inside or outside the South Pacific Nuclear Free Zone;

b) not to seek or receive any assistance in the manufacture or acquisition of any nuclear explosive device;

c) not to take any action to assist or encourage the manufacture or acquisition of any nuclear explosive device by any State.

ARTICLE 4

Peaceful Nuclear Activities

Each Party undertakes:

a) not to provide source or special fissionable material, or equipment or material especially designed or prepared for the processing, use or production of special fissionable material for peaceful purposes to:

   i) any non-nuclear-weapon State unless subject to the safeguards required by Article III.1 of the NPT, or

ii) any nuclear-weapon State unless subject to applicable safeguards agreements with the International Atomic Energy Agency (IAEA).

Any such provisions shall be in accordance with strict non-proliferation measures to provide assurance of exclusively peaceful non-explosive use;

b) to support the continued effectiveness of the international non-proliferation system based on the NPT and the IAEA safeguards system.

ARTICLE 5

Prevention of stationing of Nuclear Explosive Devices

1. Each Party undertakes to prevent in its territory the stationing of any nuclear explosive device.
2. Each Party in the exercise of its sovereign rights remains free to decide for itself whether to allow visits by foreign ships and aircraft to its ports and airfields, transit of its airspace by foreign aircraft, and navigation by foreign ships in its territorial sea or archipelagic waters in a manner not covered by the rights of innocent passage, archipelagic sea lane passage or transit passage of straits.

ARTICLE 6

Prevention of Testing of Nuclear Explosive Devices

Each Party undertakes:

a) to prevent in its territory the testing of any nuclear explosive device;

b) not to take any action to assist or encourage the testing of any nuclear explosive device by any State.

ARTICLE 7

Prevention of Dumpling

1. Each Party undertakes:

   a) not to dump radioactive wastes and other radioactive matter at sea anywhere within the South Pacific Nuclear Free Zone.

   b) to prevent the dumping of radioactive wastes and other radioactive matter by anyone in its territorial sea;

   c) not to take any action to assist or encourage the dumping by anyone of radioactive wastes and other radioactive matter at sea anywhere within the South Pacific Nuclear Free Zone;

   d) to support the conclusion as soon as possible of the proposed Convention relating to the protection of the natural resources and environment of the South Pacific region and its Protocol for the prevention of pollution of the South Pacific region by dumping, with the aim of precluding dumping at sea of radioactive wastes and other radioactive matter by anyone anywhere in the region.

2. Paragraphs l(a) and l(b) of this Article shall not apply to areas of the South Pacific Nuclear Free Zone in respect of which such a Convention and Protocol have entered into force.

ARTICLE 8

Cpntrol System

1. The Parties hereby establish a control system for the purpose of verifying compliance with their obligations under this Treaty.

2. The control system shall comprise:
   a) reports and exchange of information as provided for in Article 9;
   b) consultations as provided for in Article 10 and Annex 4(1);
   c) the application to peaceful nuclear activities of safeguards by the IAEA as provided for in Annex 2;
   d) a complaints procedure as provided for in Annex 4.

ARTICLE 9

Reports of Exchanges of Information

1. Each Party shall report to the Director of the South Pacific Bureau for Economic Co-operation (the Director) as soon as possible any significant event within its jurisdiction affecting the implementation of this Treaty. The Director shall circulate such reports promptly to all Parties.
2. The Parties shall endeavour to keep each other informed on matters arising under or in relation to this Treaty. They may exchange information by communicating it to the Director, who shall circulate it to all Parties.
3. The Director shall report annually to the South Pacific Forum on the status of this Treaty and matters arising under or in relation to it, incorporating reports and communications made under paragraphs 1 and 2 of this Article and matters arising under Articles 8(2)(d) and 10 and Annex 2(4).

ARTICLE 10

Consultations and Review

Without prejudice to the conduct of consultations among Parties by other means, the Director, at the request of any

Party, shall convene a meeting of the Consultative Committee established by Annex 3 for consultation and co-operation on any matter arising in relation to this Treaty or for reviewing its operation.

ARTICLE 11

Amendment

The Consultative Committee shall consider proposals for amendment of the provisions of this Treaty proposed by any Party and circulated by the Director to all Parties not less than three months prior to the convening of the Consultative Committee for this purpose. Any proposal agreed upon by consensus by the Consultative Committee shall be communicated to the Director who shall circulate it for acceptance to all Parties. An amendment shall enter into force thirty days after receipt by the depositary of acceptances from all Parties.

ARTICLE 12

Signature and Ratification

1. This Treaty shall be open for signature by any Member of the South Pacific Forum.
2. This Treaty shall be subject to ratification. Instruments of ratification shall be deposited with the Director who is hereby designated depositary of this Treaty and its Protocols.
3. If a Member of the South Pacific Forum whose territory is outside the South Pacific Nuclear Free Zone becomes a Party to this Treaty, Annex 1 shall be deemed to be amended so far as is required to enclose at least the territory of that Party within the boundaries of the South

Pacific Nuclear Free Zone. The delineation of any area added pursuant to this paragraph shall be approved by the South Pacific Forum.

ARTICLE 13

Withdrawal

1. This Treaty is of a permanent nature and shall remain in force indefinitely, provided that in the event of a violation by any Party of a provision of this Treaty essential to the achievement of the objectives of the Treaty or of the spirit of the Treaty, every other Party shall have the right to withdraw from the Treaty.
2. Withdrawal shall be effected by giving notice twelve months in advance to the Director who shall circulate such notice to all other Parties.

ARTICLE 14

Reservations

This Treaty shall not be subject to reservations.

ARTICLE 15

Entry into Force

1. This Treaty shall enter into force on the date of deposit of the eighth instrument of ratification.
2. For a signatory which ratifies this Treaty after the date of deposit of the eighth instrument of ratification, the Treaty shall enter into force on the date of deposit of its instrument of ratification.

ARTICLE 16

Depository Functions

The depositary shall register this Treaty and its Protocols pursuant to Article 102 of the Charter of the United Nations and shall transmit certified copies of the Treaty and its Protocols to all Members of the South Pacific Forum and all States eligible to become Party to the Protocols to the Treaty and shall notify them of signatures and ratifications of the Treaty and its Protocols.

IN WITNESS WHEREOF the undersigned, being duly authorized by their Governments, have signed this Treaty.

DONE at Rarotonga, this sixth day of August, One thousand nine hundred and eighty-five, in a single original in the English language.

ANNEX 1

South Pacific Nuclear Free Zone

A. The area bounded by a line:

1) commencing at the point of intersection of the Equator by the maritime boundary between Indonesia and Papua New Guinea;

2) running thence northerly along that maritime boundary to its intersection by the outer limit of the exclusive economic zone of Papua New Guinea;

3) thence generally north-easterly, easterly and southeasterly along that outer limit to its intersection by the Equator;

4) thence east along the Equator to its intersection by the meridian of Longitude 163 degrees East;

5) thence north along that meridian to its intersection by the parallel of Latitude 3 degrees North;

6) thence east along that parallel to its intersection by the meridian of Longitude 171 degrees East;

7) thence north along that meridian to its intersection by the parallel of Latitude 4 degrees North;

8) thence east along that parallel to its intersection by the meridian of Longitude 180 degrees East;

9) thence south along that meridian to its intersection by the Equator;

10) thence east along the Equator to its intersection by the meridian of Longitude 165 degrees West;

11) thence north along that meridian to its intersection by the parallel of Latitude 5 degrees 30 minutes North;

12) thence east along that parallel to its intersection by the meridian of Longitude 154 degrees West;

13) thence south along that meridian to its intersection by the Equator;

14) thence east along the Equator to its intersection by the meridian of Longitude 115 degrees West;

15) thence south along that meridian to its intersection by the parallel of Latitude 60 degrees South;

16) thence west along that parallel to its intersection by the meridian of Longitude 115 degrees East;

17) thence north along that meridian to its southernmost intersection by the outer limit of the territorial sea of Australia;

18) thence generally northerly and easterly along the outer limit of the territorial sea of Australia to its intersection by the meridian of Longitude 136 degrees 45 minutes East;

19) thence north-easterly along the geodesic to the point of Latitude 10 degrees 50 minutes South, Longitude 139 degrees 12 minutes East;

20) thence north-easterly along the maritime boundary between Indonesia and Papua New Guinea to where it joins the land border between those two countries;

21) thence generally northerly along that land border to where it joins the maritime boundary between Indonesia and Papua New Guinea, on the northern coastline of Papua New Guinea; and

22) thence generally northerly along that boundary to the point of commencement.

B. The areas within the outer limits of the territorial seas of all Australian islands lying westward of the area described in paragraph A and north of Latitude 60 degrees South, provided that any such areas shall cease to be part of the South Pacific Nuclear Free Zone upon receipt by the depositary of written notice from the Government of Australia stating that the areas have become subject to another treaty having an object and purpose substantially the same as that of this Treaty.

ANNEX 2

IAEA Safeguards

1. The safeguards referred to in Article 8 shall in respect of each Party be applied by the IAEA as set forth in an agreement negotiated and concluded with the IAEA on all source or special fissionable material in all peaceful nuclear activities within the territory of the Party, under its jurisdiction or carried out under its control anywhere.

2. The agreement referred to in paragraph 1 shall be, or shall be equivalent in its scope and effect to, an agreement required in connection with the NPT on the basis of the material reproduced in document INFCIRC/153 (Corrected) of the IAEA. Each Party shall take all appropriate steps to ensure that such an agreement is in

force for it not later than 18 months after the date of entry into force for that Party of this Treaty.

3. For the purposes of this Treaty, the safeguards referred to in paragraph 1 shall have as their purpose the verification of the non-diversion of nuclear material from peaceful nuclear activities to nuclear explosive devices.
4. Each Party agrees upon the request of any other Party to transmit to that Party and to the Director for the information of all Parties a copy of the overall conclusions of the most recent report by the IAEA on its inspection activities in the territory of the Party concerned, and to advise the Director promptly of any subsequent findings of the Board of Governors of the IAEA in relation to those conclusions for the information of all Parties.

ANNEX 3

Consultative Committee

1. There is hereby established a Consultative Committee which shall be convened by the Director from time to time pursuant to Articles 10 and 11 and Annex 4 (2). The Consultative Committee shall be constituted of representatives of the Parties, each Party being entitled to appoint one representative who may be accompanied by advisers. Unless otherwise agreed, the Consultative Committee shall be chaired at any given meeting by the representative of the Party which last hosted the meeting of Heads of Government of Members of the South Pacific Forum. A quorum shall be constituted by representatives of half the Parties. Subject to the provisions of Article 11, decisions of the Consultative Committee shall be taken by consensus or, failing consensus, by a two-thirds majority of those present and voting. The Consultative Committee shall, adopt such other rules of procedure as it sees fit.

2. The costs of the Consultative Committee, including the costs of special inspections pursuant to Annex 4, shall be borne by the South Pacific Bureau for Economic Co-operation. It may seek special funding should this be required.

ANNEX 4

Complaints Procedure

1. A Party which considers that there are grounds for a complaint that another Party is in breach of its obligations under this Treaty shall, before bringing such a complaint to the Director, bring the subject matter of the complaint to the attention of the Party complained of and shall allow the latter reasonable opportunity to provide it with an explanation and to resolve the matter.
2. If the matter is not so resolved, the complainant Party may bring the complaint to the Director with a request that the Consultative Committee be convened to consider it. Complaints shall be supported by an account of evidence of breach of obligations known to the complainant Party. Upon receipt of a complaint the Director shall convene the Consultative Committee as quickly as possible to consider it.
3. The Consultative Committee, taking account of efforts made under paragraph 1, shall afford the Party complained of a reasonable opportunity to provide it with an explanation of the matter.
4. If, after considering any explanation given to it by the representatives of the Party complained of, the Consultative Committee decides that there is sufficient substance in the complaint to warrant a special inspection in the territory of that Party or elsewhere, the Consultative Committee shall direct that such special

inspection be made as quickly as possible by a special inspection team of three suitably qualified special inspectors appointed by the Consultative Committee in consultation with the complained of and complainant Parties, provided that no national of either Party shall serve on the special inspection team. If so requested by the Party complained of, the special inspection team shall be accompanied by representatives of that Party. Neither the right of consultation on the appointment of special inspectors, nor the right to accompany special inspectors, shall delay the work of the special inspection team.

5. In making a special inspection, special inspectors shall be subject to the direction only of the Consultative Committee and shall comply with such directives concerning tasks, objectives, confidentiality and procedures as may be decided upon by it. Directives shall take account of the legitimate interests of the Party complained of in complying with its other international obligations and commitments and shall not duplicate safeguards procedures to be undertaken by the IAEA pursuant to agreements referred to in Annex 2 (1). The special inspectors shall discharge their duties with due respect for the laws of the Party complained of.
6. Each Party shall give to special inspectors full and free access to all information and places within its territory which may be relevant to enable the special inspectors to implement the directives given to them by the Consultative Committee.
7. The Party complained of shall take all appropriate steps to facilitate the special inspection, and shall grant to special inspectors privileges and immunities necessary for the performance of their functions, including inviolability for all papers and documents and immunity from arrest, detention and legal process for acts done and words spoken and written, for the purpose of the special inspection.

8. The special inspectors shall report in writing as quickly as possible to the Consultative Committee, outlining their activities, setting out relevant facts and information as ascertained by them, with supporting evidence and documentation as appropriate, and stating their conclusions. The Consultative Committee shall report fully to all Members of the South Pacific Forum, giving its decision as to whether the Party complained of is in breach of its obligations under this Treaty.
9. If the Consultative Committee has decided that the Party complained of is in breach of its obligations under this Treaty, or that the above provisions have not been complied with, or at any time at the request of either the complainant or complained of Party, the Parties shall meet promptly at a meeting of the South Pacific Forum.

On 8 August 1986 the Protocols to the South Pacific Nuclear Free Zone Treaty were adopted by the South Pacific Forum at its 17th session, in Suva.

The following texts of the Protocols were formally communicated to the Director General by the Director of the South Pacific Bureau for Economic Co-operation (SPEC) and are herewith being circulated to all Member States for their information pursuant to a request made by the Director of SPEC.

Following the deposit of the eighth instrument of ratification, the South Pacific Nuclear Free Zone Treaty entered into force on 11 December 1986.

PROTOCOL 1

The Parties to this Protocol

Noting the South Pacific Nuclear Free Zone Treaty (the Treaty)

Have agreed as follows:

Article 1

Each Party undertakes to apply, in respect of the territories for which it is internationally responsible situated within the South Pacific Nuclear Free Zone, the prohibitions contained in Articles 3, 5 and 6, insofar as they related to the manufacture, stationing and testing of any nuclear explosive device within those territories, and the safeguards specified in Article 8(2)(c) and Annex 2 of the Treaty.

Article 2

Each Party may, by written notification to the depositary, indicate its acceptance from the date of such notification of any alteration to its obligation under this Protocol brought about by the entry into force of an amendment to the Treaty pursuant to Article 11 of the Treaty.

Article 3

This Protocol shall be open for signature by the French Republic, the United Kingdom of Great Britain and Northern Ireland and the United States of America.

Article 4

This Protocol shall be subject to ratification.

Article 5

This Protocol is of a permanent nature and shall remain in force indefinitely, provided that each Party shall, in exercising its national sovereignty, have the right to withdraw from this

Protocol if it decides that extraordinary events, related to the subject matter of this Protocol, have jeopardized its supreme interests. It shall give notice of such withdrawal to the depositary three months in advance. Such notice shall include a statement of the extraordinary events it regards as having jeopardized its supreme interests.

Article 6

This Protocol shall enter into force for each State on the date of its deposit with the depositary of its instrument of ratification.

IN WITNESS WHEREOF the undersigned, being duly authorized by their Governments, have signed this Protocol.

DONE at Suva, this eighth day of August, one thousand nine hundred and eighty-six, in a single original in the English language.

PROTOCOL 2

The Parties to this Protocol

Noting the South Pacific Nuclear Free Zone Treaty (the Treaty)

Have agreed as follows:

Article 1

Each Party undertakes not to use or threaten to use any nuclear explosive device against:

a) Parties to the Treaty; or

b) Any territory within the South Pacific Nuclear Free Zone for which a State that has become a Party to Protocol 1 is internationally responsible.

Article 2

Each Party undertakes not to contribute to any act of a Party to the Treaty which constitutes a violation of the Treaty, or to any act of another Party to a Protocol which constitutes a violation of a Protocol.

Article 3

Each Party may, by written notification to the depositary, indicate its acceptance from the date of such notification of any alteration to its obligation under this Protocol brought about by the entry into force of an amendment to the Treaty pursuant to Article 11 of the Treaty or by the extension of the South Pacific Nuclear Free Zone pursuant to Article 12(3) of the Treaty.

Article 4

This Protocol shall be open for signature by the French Republic, the People's Republic of China, the Union of Soviet Socialist Republics, the United Kingdom of Great Britain and Northern Ireland and the United States of America.

Article 5

This Protocol shall be subject to ratification.

Article 6

This Protocol is of a permanent nature and shall remain in force indefinitely, provided that each Party shall, in exercising its national sovereignty, have the right to withdraw from this Protocol if it decides that extraordinary events, related to the subject matter of this Protocol, have jeopardized its supreme interests. It shall give notice of such withdrawal to the depositary three months in advance. Such notice shall include a statement of the extraordinary events it regards as having jeopardized its supreme interests.

Article 7

This Protocol shall enter into force for each State on the date of its deposit with the depositary of its instrument of ratification.

IN WITNESS WHEREOF the undersigned, being duly authorized by their Governments, have signed this Protocol.

DONE at Suva, this eighth day of August, one thousand nine hundred and eighty-six, in a single original in the English language.

PROTOCOL 3

The Parties to this Protocol

Noting the South Pacific Nuclear Free Zone Treaty (the Treaty)

Have agreed as follows:

Article 1

Each Party undertakes not to test any nuclear explosive device anywhere within the South Pacific Nuclear Free Zone.

Article 2

Each Party may, by written notification to the depositary, indicate its acceptance from the date of such notification of any alteration to its obligation under this Protocol brought about by the entry into force of an amendment to the Treaty pursuant to Article 11 of the Treaty or by the extension of the South Pacific Nuclear Free Zone pursuant to Article 12(3) of the Treaty.

Article 3

This Protocol shall be open for signature by the French Republic, the People's Republic of China, the Union of Soviet Socialist Republics, the United Kingdom of Great Britain and Northern Ireland and the United States of America.

Article 4

This Protocol shall be subject to ratification.

Article 5

This Protocol is of a permanent nature and shall remain in force indefinitely, provided that each Party shall, in exercising its national sovereignty, have the right to withdraw from this Protocol if it decides that extraordinary events, related to the subject matter of this Protocol, have jeopardized its supreme interests. It shall give notice of such withdrawal to the depositary three months in advance. Such notice shall include a statement of the extraordinary events it regards as having jeopardized its supreme interests.

Article 6

This Protocol shall enter into force for each State on the date of its deposit with the depositary of its instrument of ratification.

IN WITNESS WHEREOF the undersigned, being duly authorized by their Governments, have signed this Protocol.

DONE at Suva, this eighth day of August, one thousand nine hundred and eighty-six, in a single original in the English language.

Bibliography

PRIMARY SOURCES

Australia and Disarmament: Steps in the Right Direction, Australia, Department of Foreign Affairs, 1986.

Australia, Parliament, Debates, House of Representatives, 23rd Parliament, 3rd Session, 15 August 1961.

Australia Parliament, Debates, "Australia's Foreign Policy", in House of Representatives (Hansard), 13 February 1986.

"Australia Welcomes UN decision on Arms Control", News Release Issued by the Minister for Foreign Affairs and Trade Senator Gareth Evans, Australian Foreign Affairs Record (AFAR), Vol. 59 No. 11, November 1988.

Address of Bill Hayden, "Australia-US Relations", *AFAR*, August 1986

AFAR, Vol. 58, No.3, March 1987, pp. 170-171

AFAR, Vol. 54, No. 10, October 1983

AFAR, Vol. 58, No.2, February 87 pp. 91&97.

AFAR, Vol. 56, No.9 September 1985, pp.865-866, 871-872

AFAR, Vol. 54, No.8, p-83

AFAR, Vol. 50, No.w February 1987, pp. 96-99

AFAR, Vol. 54, No. 8, August 1983, pp. 379-99

AFAR, Vol. 54, No.6, August 1985, p.699

AFAR, Vol. 58, No.10, 1987, November/December, pp. 647 & 653

AFAR, Vol. 58, No.7, July 87.

AFAR, Vol. 58, No.3, March 1987, pp. 149 & 159

AFAR, Vol. 57, No.6, June 1986, pp. 532-536

AFAR, Vol. 57, No.72, December 1986, pp. 1720-1721

AFAR, Vol.54, No.9, September 1983, p.538.

AFAR, Vol. 54, No.12, December 1983, pp. 790, 827

AFAR, Vol. 57, No.5, 1986, p.416

ASEAN, from Department of Foreign Affairs, ASEAN National Secretariat of Indonesia, Public Relations Division, Jakarta, June, 1975, p.65

"A World at Peace", *Report of the Palme Commission on Disarmament and Security*, Stokholm, April 1989.

Bandung Conference, May 7-14, 1955 Keesings Record of World Events, No. 10, 1955-56, pp. 14181-14183.

Cark (Helen), "Establishing a Nuclear-Free-Zone in the South Pacific", Address to the Twentieth Foreign Policy School, Otago University, New Zealand, 18 May, 1985.

Churchill's Speeches, Vol. III, Macmillan, 1959, p.281

Comprehensive Study of the Question of Nuclear-Weapon-Free Zones in All its Aspects: Special Report of the Conference of the Committee on Disarmament, (New York): UN, General Assembly Official Records 1975), 30th Session Supplement, No. 27A.

Comprehensive Study of the Question of Nuclear-Weapon Free Zones in all its Aspects, (United Nations 1976).

Disarmament and Security: 1987 Yearbook (Moscow: Novosti Press Agency Publising House, 1988).

End Paper 8, The Perugia Convention (England: Bertrand Russel Peace Foundations Ltd., 1984).

Final Document of the First UN Special Session Devoted to Disarmament, United Nations, 1978.

Final Document of the Tenth Special Session of the General Assembly (New York; United Nations 1972) A/RES/S-1012, pp. 14-15.

Final Communique of Kuala Lumpur Summit 1977 on ZOPFAN. Published in R. Negi, "*ASEAN: 20 Years*".

"Foreign Ministers' Meeting-Declaration", *Foreign Affairs Malaysia*, Vol.9, No.4, December 1971, pp. 57-59.

"French Nuclear Testing at Mururoa Atoll", New Release Issued by the Minister for Foreign Affairs and Trade Mr. Bill Hayden on May 13, 1988, *AFAR*, Vol. 59, No.5, May 1988.

"French Nuclear Testing in the Pacific", Wellington, *Ministry of Foreign Affairs*, Vol.1, 1973, No. 2, 1975.

Great Britain, Foreign Office, *Declaration and Protocol on Neutrality of Laos* (London, 1963).

Hayden, Bill, "The Review of ANZUS", Text of the statement to Parliament on 15 Sep. 1983 by the Minister for Foreign Affairs, Mr. Bill Hayden, M.P., published, in *World Review*, Vol. 23, No.4, October 1984, pp. 43-51.

"International Action on Nuclear Safety: Australian Proposal", News Release Issued by Minister for Foreign Affairs Mr. Bill Hayden, *AFAR*, Vol.57, No.5, May 1986.

International Meeting For Nuclear Weapon Free Zones (A Documentary Account), Berlin, 20-22 June 1988, Taken from Pakistan Horizon, Vol.43, No.1, 1990, p.67.

James, Baker, "A Framework for Future Pacific Partnership", *Address to the Asia Society*, New York, June 26, 1989.

Nehru, Jawaharlal, *India's Foreign Policy* (Selected Speeches September 1946-April 1961), New Delhi, 1961, p.67.

Joint Communique of the 20th ASEAN Ministerial Meeting in Singapore on July 15-16, 1987. Published in R. Negi, *ASEAN: 20 Years.*

Joint Communique, ANZUS Defence Ministry Meeting, Wellington, 3 April 1985.

Joint Press Statement of ASEAN, Heads of Govt., Manila, December 1987, Published in R. Negi, *ASEAN: 20 Years.*

Goldblat, Joseph, "Arms Control agreement", *SIPRI*, (Stockholm 1983).

Keesing's Contemporary Archives: *Record of World Events*, Longman, UK.

Lange, David, "New Zealand's Security Policy", *Foreign Affairs*, Vol. 63, No.5, 1985.

_______________, *Address Delivered to the Institute of International Affairs and the National Press Club*, Wellington, November 27, 1985.

Soon, Lau Teik, *Indonesia and the Regional Security, The Djakarta Conference on Combodia* (Singapore, 1972), p.6

Mahhubani (Kishore), then Deputy Chief Mission of the Singapore Embassy in Washington, "The Kampuchian Problem: A Southeast Asian Perception", *Foreign Affairs*, Vol. 62, No.2, Winter 1983/84.

Malaysia Information Bulletin (Malaysian High Commission, New Delhi), 21 April 1971.

Mikhail Gorbachev in his address to the UN General Assembly, on December 7, 1988.

Neutralisation in South East Asia: Problems and Prospects, (US Government printing office, Washington, 1976), p. III.

New Zealand Ministry of Foreign Affairs, "Disarmament and Arms Control", *Information Bulletin*, No.3, October 1982.

New Zealand Official Yearbook, Department of Statistics, Wellington.

"Nuclear Disarmament and Arms Control: Visit by Soviet Official". News release issued by the Minister for Foreign Affairs, Mr. Bill Hayden, M.P. on 17 January,*AFAR*, Vol. 57, No.1. January 1986.

Nuclear Arms Control: Background and Issues, Committee on International Security and Arms Control, National Academy of Sciences (National Academy Press, Washington DC., 1985).

Nuclear-Free Zone and Nordic Security: A Special Issue of Ulkipolitickka, Helsinki 1975, Finish Institute of International Affairs.

Official Yearbook of the Common Wealth of Australia (Annual).

Report of the Committee to Review Australia's Relations with the Third World, Australian Government Publishing Service, Canberra, 1979.

Report of the Joint Committee on Foreign Affairs and Defence, "Australia and ASEAN: Challenges and Opportunities", Canberra: Australian Government Publishing Service, 1984.

Report of the Working Group submitted by its Chairman David Sadleir on a South Pacific Nuclear Free Zone to the South Pacific Forum, Rarotonga, August 4-6, 1985.

Report by the Chairman of the Working Group on a South pacific Nuclear Free Zone to the South Pacific Forum, *South Pacific Forum*, Suva, 1985.

Resolution No. 3477 (xxx): Establishment of a nuclear-weapon-free zone in the South Pacific, 13th Session of UN, General Assembly, 11 December 1975.

Shultz George, P., "On Alliance Responsibility", address given at the East-West Centre Honululu, Hawaii, July 17, 1985, *Current Policy*, 724 (Washington D.C.: Dept. of State, 1985).

Security Treaty Between Australia New Zealand and the United States, (Sub Committee on Asian and Pacific Affairs, Committee on Foreign Affairs, 99th Cong. 1st sess. March 18, 1985.

Select Documents on International Affairs, Department of External Affairs, Canberra, Australia .

Shadow Minister for Foreign Affairs Mackellar (Micheal), "The Govt. and International Security Critique", *World Review*, Vol. 23, No.4, October 84, pp. 52-72.

Shafie, M. Ghaizalbin, The Neutralization of Southeast Asia, *Pacific Community*, Vol.2, No.2, 1971.

"South East Asia: The Neutralization proposal", *Current Notes on International Affairs*, Vol. 44, No.10, October 1972, p.498.

Speech made by Australian Minister for Foreign Affairs, Bill Hayden on 21 May 1986, at the University of the South Pacific, Fiji, *Text in Australian Foreign Affairs Record*, Vol. 57, No.65, 1986.

Speech by Sir Wallace Rouling, Former Prime Minister of New Zealand, at the Dundui Branch of New Zealand Institute of International Affairs, 22nd March, 1984.

Speech of Hon. Russel Marshall Minister of Foreign Affairs on Nuclear Free Zone 15th March 1989, New Zealand High Commission Publication, No.2, 1990.

Speech of Hon. Russel Marshall Minister of Foreign Affairs on Nuclear Free Zone 15th March 1989, New Zealand High Commission, Publication, No.20, 1990.

Speech by Australian Prime Minister, Bob Hawke, in the *Earle Page Collage*, University of New England, Armidale, on 11 September 1986.

Speech by Australian Prime Minister the Hon. R.J. Hawke, to the Washington Press Club, Washington, USA, 15th June 1983. Reported in Backgrounder, 22nd June 1983. Annexure I.

Statements and News Conferences of Secretary of State Shultz's trip to the Philippines, Singapore, Australia and Western Samoa, *US Department of State Bulletin*, 1987.

Statements of Foreign Policy, Foreign Minister Andrew Peacock's Press Interview on March 1976, Department of Foreign Affairs, Canberra, 1976.

Statement by Australian Foreign Minister Bill Hayden in Parliament, 5 June 1986.

Statement made by Fillipe Bole, Fiji's Permanent Representative to the UN, to the General Assembly, 13 June 1982. Also at the 35 the Session 10th Oct 1980.

Statement by General Soeharto, President of the republic of Indonesia at IInd Summit in Kuala Lumpur 1971, published in R. Negi, *ASEAN*: 20 Years.

Study on All the Aspects of Regional Disarmament (New York: United Nations 1981). A/35/416, pp. 14-19.

Text of the Antartic Treaty, Published in *SIPRI Yearbook*, 1973, pp.487-493.

Text of Treaty the Progibition of the Emplacement of Nuclear Weapons and other Weapons of Mass Destruction on the Sea-Bed and the Ocean Floor and in the Subsoil thereof, Published in *SIPRI* 1972, pp. 537-5411.

Text of British Government Announcement, issued on 20th March 1987.

Text of the Treaty for the Prohibition of Nuclear Weapons in Latin America (Treaty of Tlatelolco) published in *SIPRI Year Book of World Armaments and Disarmament* 1969/70. (New Delhi: Oxford 4 IBH Pub. Co. 1970), pp. 237-255.

The New Zealand Nuclear Free Zone, Disarmament and Arms Control Act Passed on 5th June 1987.

The ASEAN Declaration, Bangkok, August 8, 1967, taken from R. Negi, *ASEAN: 20 years* (New Delhi: Lancers Books, 1989).

The South Pacific Forum, *Information Bulletin No. 26*, July 1989. Ministry of External Relations and Trade, New Zealand.

The United Nations and Disarmament: 1945-1985, New York: United Nations (1985).

The Far East and Australia Yearbook, Europa Pub. Ltd.

The United Nations and Disarmament 1970-1975 (New York: United Nation, 1976).

'The Second NPT Review Conference, *SIPRI Yearbook*, 1981, p.330.

Treaty on the Non-Proliferation of Nuclear Weapons, News Release Issued by Prime Minister Mr. Bob Hawke on July 1 1988, *AFAR*, Vol. 59, No.7 July 1988.

United Nations Centre for Disarmament, *Study on all the Aspects of Regional Disarmament*, New York, UN, 1981, Study Series 3. *Nuclear Weapons: A Comprehensive Study*, New York, 1991, Study Series 21.

United Nations Convention on the Law of the Sea, Article 210 (Pollution by Dumping).

US House of Representatives, *The South Pacific Nuclear Free Zone, Hearings and Markup before the Committee of Foreign Affairs and its Sub-Committee on Asian and Pacific Affairs.*

US Arms Control & Disarmament Agency, *Arms Control and Disarmament Agreements: Texts and History of the Negotiations*, 1990 (Washington: USGPO 1990).

The Theory of NWFZs and WMDFZs, United Nations,1996, pp 35-57.

"Regional Disarmament", *The UN Disarmament Yearbook,2000,* pp 139-163.

SECONDARY SOURCES

Books

Acharya, Amitav, *The Quest for Identity: International Relations of Southeast Asia* (Oxford University Press, 2000).

Cunha, Derek da (ed.), *Southeast Asian Perspective on Security* (Singapore: Institute of Southeast Asian studies, 2000).

Booth, (Ken) and Trood (Russell), *Strategic Cultures in the Asia-Pacific Region* (London: Macmillian Press Ltd., 1999).

Leifer, Michael, *ASEAN and the Security: South-East Asia* (London and New York: Routledge, 1989).

Singh, Bilveer, *ZOPFAN and the New Security Order in the Asia-Pacific Region* (Malaysia: Pelanduk, Publications, 1992).

Palmer, Ronald D. and Reckford (Thimas J), *Building Asean: 20 Years of Southeast Asian Cooperation* (New York, 1984).

Andrews, E.M., *A History and Australian Foreign Policy: From Dependence to Independence* (Longman Cheshire, 1979).

Simone, Vera and Ferauru (Anne Thompson), *The Asian Pacific: Political and Economic Development in a Global Context* (Longman, 1995).

Malik, J. Mohan, *Australia's Security in the 21st Century* (Australia: Allen & Unwin 1999).

Fry, Greg, *Australia's Regional Security* (Australia: Allen & Unwin,1991).

Cheeseman, Gracme and Bruce (Robert), *Discourse and Danger & Dread Frontiers: Australian Defence and Security Thinking After the Cold War* (Australia, 1996).

Chakrabarthy, H.R., *Vietnam, Kampuchia, Laos, Bound In Comradership* (New Delhi: Patriot Publishers, 1988).

Kaiser, David, *American Tragedy* (London: The Belknap Press of Harvard University Press, 2000).

Firth, Stewart, *Nuclear Playground* (London: Allen & Unwin, 1987).

Abdulgani, Roeslan, *The Bandung Connection: The Asia-Africa Conference in Bandung in 1955* (Singapore MCM-LXXXI: Gunnung Agung, 1981).

——, *Nationalism, Regionalism and Security: Problems in South East Asia*, (New Delhi: Banyan Pub., 1984).

Alagappa, Muthiah, *US-ASEAN Security: Limits and Possibilities*, (Institute of Strategic and International Studies (ISIS), Malaysia, 1986).

Alley Roderic, *Nuclear-Weapon Free Zones: The South Pacific Proposal* (Muscatine, Iowa: The Stanley Foundation, 1977).

Barnaby, Frank and Thomas Geoffrey eds., *Nuclear Arms Race: Control or Catastrophe* (London: Frances Printer Publication 1982).

Bebler Ales, *National Sovereignty in the Nuclear Age* (New Delhi: Indian Council for Cultural Relations, 1972).

Bhattacharjee G.P., *South-East Asian Politics: Malaysia and Indonesia*; (Calcutta: Minerva 1976).

Butwell, Richard, *South East Asia: A Political Introduction* (New York: Praeger Pub., 1975).

Carter A., *Peace Movements: International Protests and World Politics Since 1945* (London: Longman, 1992).

Chawla, Sudershan and others (ed.), *South-East Asia under the New Balance of Power* (New York: Praeger, 1974).

——., "US Strategy in Southeast Asia in the Post-Cease-Fire Period", *Southeast Asia Under the New Balance of Power* (New York: Praeger, 1974).

Colbert, Evelya, *South-East Asia in International Politics 1941-56* (Ithaca; Cornell University Press, 1977).

Coox, Alvin D., "Japan And the Southeast Asian Cinfiguration", *Southeast Asia Unver New Balance of Power* (New York: Praeger, 1974).

Dahlitz Julie, *Nuclear Arms Control* (London: George Allen and Unwin, 1983).

Das, Parimal Kumar (ed.), *Troubled Region: Issues of Peace and Development in South East Asia* (New Delhi: Sage Pub., 1987).

Dewitt, David B., (ed.), *Nuclear Non-Proliferation and Global Security* (London: Croom Helm, 1987).

Dahl, Robert, *Controlling Nuclear Weapons: Democracy Versus Guardianship* (Syracuse USA: Syracuse University Press, 1985).

Dorrance, John C., *Oceania and the US: An Analysis of the US Interests and Policy in the South Pacific* (Washington DC, 1980).

Drell, Sidney D., *Facing the Threat of Nuclear Weapons*, (University of Washington Press, Seattle and London, 1983).

Dutt, Ashok K., *Southeast Asia: Realm of Contrasts*, (Boulder Westview Press 1985).

Epstein, William, *The Last Chance: Nuclear Proliferation and Arms Control* (New York: Free Press, 1985).

Finis, John and others, *Nuclear Deterrence: Morality and Realism* (Oxford: Clarendon Press, 1987).

Fischer, George (Translated by David Willey), *The Non-Proliferation of Nuclear Weapons* (London: Europa Publications, 1970).

——*Southeast Asia: A Social, Economic and Political Geography* (London: Metheun, 1964).

Human Rights and the Basic Needs in the American (Washington DC: Georgetown University Press), 1982.

Isreal, Fred L. (ed.), *Major Peace Treaties of Modern History 1648-1967* (New York: Chelsea House, 1967).

Jukes Geoffery, *Strategic Situation in the 1980: An Exercise in Forecasting* (Canberra: Australian National University Press, 1968).

K. Subrahmanyam (ed.), *Nuclear Proliferation and International Security* (New Delhi: IDSA, 1985).

——, (ed.), *Nuclear Myths and Realities*, (New Delhi: 1980).

Kaul, Man Mohini, *Pearls in the Ocean: Security Perspectives in South-West Pacific* (New Delhi: UBSPD, 1993).

Khan A.R., & Hafiz M.A., *Security of Small States* (Dhaka: University Press, 1987).

Mahapatra, Chintamani, *American Role in Origin and Growth of ASEAN* (New Delhi: ABC Publishing House, 1990).

Marsot, Alian-Gerard, The Chinese Perspective in *Southeast Asia under New Balance of Power* (New York: Praeger, 1974).

Mcalister, John T., (ed.), *South East Asia: The Politics of National Integration* (New York: Random House, 1973).

Menons, Dennis, *World at Risk: The Debate over Arms Control* (Jefferson, North Carolina and London: McFarland and Company, Inc. Pub., 1986).

Mishra, K.P., and V.D. (Chopra) (eds.), *South-east Asia Pacific Region: Emerging Trends*, (New Delhi: International Institute of Southern Asia-Pacific Studies, 1988)

——Rasgotra M. and Chopra V.D., *Southern Asia Pacific: Perceptions and Strategies* (New Delhi: Continental Publishing House, 1988)

Negi, R., *ASEAN-20 Years* (New Delhi: Lancers Books, 1989).

Nehru, Jawaharlal, *India's Foreign Policy* (Selected Speeches, September 1946- April 1961, (New Delhi: 1961).

Osbern, Milton, *Southeast Asia: An Introductory History* (Sydney: Allen and Unwin, 1984).

Park, Jae Lyu, (ed.), *Nuclear Proliferation in Developing Countries* (Seoul: Institute of Far Eastern Studies, Kyungnam University, 1979).

Pathak, K.K.,*Nuclear Policy of India: A Third World Perspective*, (New Delhi: Gitanjali Prakashan, 1980).

Poulose, T.T., *Nuclear Proliferation and the Third World* (New Delhi: ABC Pub, 1982).

Robles, Mlfonso Garcia, *Nuclear Disarmament: A Crucial Issue for the Survival of Mankind* (New Delhi:ICCR, 1984).

Sardesai, D.R., *South East Asia: Past and Present* (New Delhi: Vikas Pub. 1981).

Shere, Waris, ed., *In Search of Peace* (Karachi: Royal Book Company, 1988).

Sherman, Michael E., *Nuclear Proliferation: The Treaty And After* (Canadian Institute of International Affairs, 1968).

Simpson, John (ed.), *Nuclear Non-Proliferation: An Agenda for the 1990s* (Cambridge: Cambridge University Press, 1987).

"South-East Asian Nuclear Weapon Free Zone: Ideas Problems, Prospects and Proposals", *Regional Disarmament Workshop for Asia Pacific*, (New York: UN, 1991.)

Southeast Asia under the New Balance of Power, Praeger Special Studies in International Polities & Government (New York: Praeger Publishers, 1974).

Stephen, John (ed.), *The Soviet-US Horizon on the Pacific*, (Honolulu: University Hawaii Press, 1987).

Subrahmanyam, K. (ed.),*Nuclear Myths and Realities: India's Dilemma* (New Delhi: ABC Pub. 1981).

The Pacific Community: An Outlook (Moscow: Progress Pub, 1987).

The Far East and Australia Yearbook (London: Europae Pub. Ltd.).

Williams, Phil (ed.), *Nuclear Debate: Issues and Politics* (London: Routledge and Kegan Paul for the Royal Institute of International Affairs, 1984).

Articles

Abdus Sabur, A.K.M, "quest for a viable Regional Order in South East Asia: Problems and Prospects",*Biiss Journal*, Vol.7 No.4, October 1986, pp. 475-506.

Aganzhanov, S., "Southeast Asia and the Cambodian Problem", *Socialism: Theory and Practice*, Vol.6, No. 203, June 1990, pp. 109-14.

Agyeman-Duah, Baffour, "Nuclear Weapon Free Zones and Disarmament",*Africa Today*, Vol. 32, No. 1&2, 1985, pp. 71-90.

Alagappa, Muthiah, "Dynamics of International Security in Southeast Asia: Change and Continuity", *Australian Journal of International Affairs*, Vol.45, No.7, May 91, pp.1-37.

______________ , "Security in South-East Asia: Beyond a Zone of Peace, Freedom and Neutrality (ZOPFAN)", *Disarmament: Topical Papers 6*, (New York: UN, 1991).

______________ "A Nuclear Weapons Free Zone in Southeast Asia: Problem and Prospects", *Australian Outlook*, Vol. 41, No.3, December 1987.

______________ "Regional Arrangements and International Security in South East Asia: Going Beyond ZOPFAN", *Contemporary Southeast Asia*, Vol. 12, No.4, March 1991, pp. 269-305.

______________ "Soviet Policy in Southeast Asia: Towards Constructive Egnagement", *Pacific Affairs*, Vol. 63, No.3, Fall 1990, pp. 321-50.

______________ "Regionalism And the Quest for Security: ASEAN and the Cambodian Conflict", *Australian Journal of International Affairs*, Vol. 47, No.2, October 93, pp. 189-209.

______________, Regionalism and the Quest for Security : ASEAN and the Cambodian Conflict, *Journal of International Affairs*, Vol. 46, No.2, winter 1988, pp. 439-68.

Alves, Dora, "US-New Zealand Relations: The National Government of New Zealand",*Asian Survey*, Vol. 31, No. 11, November 91, pp. 1061-78.

"Anti-nuclear movement in the South Pacific", *Bulleting of Concerned Scholars*, Vol. 18, No.2, April -June 1986, p.89.

Armstrong (George), "An Ideology and Theology of an Independent and Nuclear Free Pacific", *Social Alternatives*, Vol.5, No.1, 1985, pp.12-13.

Arora, B.D, "Cambodia: Peace at Last", *Link*, Vol. 34, No.5, 15 Sep. 91 pp. 23-25.

Arshad, (Col. C.M), "Southeast Asia in Soviet Perspective," *Asian Defence Journal*, 5/85.

"ASEAN and the South Pacific Forum and the Changing Stratgegic Environment", *Indonesian Quarterly*, Vol.28, No.2, 1992, pp. 207-18.

ASEAN and the South Pacific", *Indonesian Quarterly*, Vol. 19, No.4, 1991, pp. 300-99 (series of articles).

"Asia", *Current History*, Vol. 89, No. 545, Mar. 1990 pp. 97-124.

Asian Survey, Vol.29, No.8, August 1989, pp. 800-815

"Australia-Indonesia Relation: Facing in 21st Century", *Indonesian Quarterly*, Vol. 20, No.2, 1992, pp. 142-218.

"Australia-New Zealand Nuclear Test Ban Resolution", *Australian Foreign Affairs Record*, Vol. 55, No. 11, November 1984, pp. 1180-82.

" A World of More Nuclear Armed Nations", *Adelphi Papers*, No. 263, winter 91.

Ayson, Robert, "Towards a Nuclear Weapon Free World?" *International Journal*, Autumn 2000, pp.531-543.

"Ballistic Missile Proliferation", *Orbis*, Vol. 35, No.1, Winter 1991, pp. 5-40.

Banerjee, D., "External Powers' Interests and Role in South-east Asia", *Strategic Analysis*, Vol. 11, No.11, February 1988, pp. 1279-92.

________________, "Southeast Asia Security in the last decade of the twentieth century", *Strategic Analysis*, Vol. 13, No.4, July 1989, pp. 365-91.

________________, " Nuclear Disarmament in the Asia Pacific Region", *Strategic Analysis*, Vol. 12,68, November 89, 803-14.

Beri, H.M.L., "security Problems of South East Asia", *IDSA Journal*, Vol. 17, No.3, January-March, 1985, pp. 345-76.

Bhardwaj, Ramdev, "China-Indonesia Relations: Towards a new era; *Strategic Analysis*, Vol. 15, No. 5, August 1992, pp.401-16.

________________, "Indian Ocean: Towards a Zone of Peace", *Monthly Public Opinion Surveys*, Vol. 36, No.1-3, Oct.November 1990, pp. 7-11.

Bhattacharya, S.S., "Indian Ocean as a zone of peace?" *Strategic Analysis*, Vol. 15, No.5, August 1992, pp. 417-40.

Biddick, Thomas V., "Diplomatic Rivalry in South Pacific: The PRC and Taiwan", *Asian Survey*, Vol. 29, No.8, August 1989, pp. 800-15.

Bidwai, Praful "Nuclear Weapons Free Zone", *Seminar*, No. 324, August 1986, pp. 27-33.

Blechman, Barry M, Mark R. Moose, "Nuclear Weapon Free Zone in Europe", *Scientific American*, Vol. 24, No.4, April 1987, pp. 37-42.

Boundreau, Donald G., "On advancing Non-Proliferation", *Strategic Analysis*, Vol. 14, No. 11, Feb 1992, pp. 1327-42.

Brown, Frederick Z., "Security issues in Southeast Asia: Proceedings of the Acacemy of Political Sciences", Vol. 38, No.2, 1991, pp. 120-30.

Bukaranbe, Bukar, "Zone of Peace or Strategic Primary: Politics of Security in the Indian Ocean",*Bulletin of Peace Proposals*, Vol. 16, No.1, Mar. 1985, pp. 51-60.

Buszynski, (Leszek), "Asean's New Challenges",*Pacific Affairs*,Vol.70, No.-4, Winter 1997-98.

————, "United States and Southeast Asia: A case of Strategic Surrender", *Journal of Southeast Asian Studies*, Vol. 14, No.2, September 83, pp. 225-43.

————, "Southeast Asia in the Post-cold War: Regionalism and Security", *Asian Survey*, Vol. 32, No.9, September 1992, pp. 830-97.

Buzan, Barry, "South East Asian Security Complex", *Contemporary South East Asia*, Vol. 10, No. 1, June 1988, pp. 1-6.

Carasales, Jalio Cesar, "South Pacific Nuclear Free Zone: Rarotonga on the steps of Tlatelolco", *World Review*, Vol. 26, No.4, Dec. 1987. pp. 53-76.

Chang, Kuang, "China's Claim of Soveregnity over Spratley and Paracel Islands: A Historical and Legal Prespective", *Case Western Reserve Journal of International Law*, Vol. 23, No. 3, summer 1991.

Chesneaux, Jean, "France in the Pacific: Global Approach or Respect for Regional Agendas",*Bulletin of Concerned Asian Scholars*, Vol. 18, No. 2 April-June 86, pp. 73-80.

Ching, Frank, "Russia Wants Active Role in Asia", *Far Eastern Economic Review*, Vol. 63, No. 32, August 10, 2000.

"Chinese in Southeast Asia", *Solidarity*, Vol. 123, July-September 1989, pp.96-125.

Chopra, V.D., "Sino-US Relations and Kampuchea", *Contemporary Affairs*, Vol.2, No.1, Jan. Mar. 88, pp. 26-36.

Choudhary, Nilufar, "Indian Ocean: A Zone of Peace or a Zone of Conflict?"*Biss Journal*, Vol. 6, No.2, April 1985, pp. 246-67.

Clark, Helen., "Establishing a Nuclear Free Zone in the South Pacific", *Gold*, 1985, pp. 121-26.

Clements, Kewin C., "New Zealand's Role in Promoting a Nuclear Free Pacific",*Journal of Peace Research*, Vol. 25, No 4, December 1988, pp. 395-410.

Cole Rodney V., "Economic Development in the South Pacific: Promoting the Private Sector", *World Development*, Vol. 2, No.2, February 1993, pp. 223-46.

Daniuelsson, Bengt, "Poisoned Pacific: The Legacy of French Nuclear Testing", *Bulleting of the Atomic Scientists*, Vol. 46, No.2, March 90, pp.

Datt, Savita, "NPT and the Non-Nuclear Weapon States: Options and Non-options", *Strategic Analysis*, Vol. 15, No. 10, January 93, pp. 911-24.

Deng, Yong, "Sino-Thai Relations form Strategic Co-operation to Economic Diplomacy", *Contemporary Southeast Asia*, Vol. 13, No. 4, March 1992, pp. 360-74.

Desmond, Edward W., "Rivalry in the Pacific: New Focus for Super Power Competition", *Time*, 24 November , 1986, pp. 4-8.

Denoon, David B.H., and Colbert, Evelyn, "Challenges for the Association of South east Asian Nations," *Pacific Affairs*, Vol. 71, No.-4, Winter 1998-99.

Dewitt (David) and Bow (Brian), "Proliferation Management in Southeast Asia," *Survival*, Vol. 38, No.3, Autumn 1996, pp. 67. 81.

Djiwandono, J. Soedjati, "ASEAN and the US Bases", *Indonesian Quarterly*, Vol. 16, No.2, April 88, pp. 161-165.

Dhamjal, Gursharan S, "Nuclear Free Zone: A Threat to Indian Security", *Link*, Vol. 34, No.37, 26 April 1992, pp. 2-28.

Din Allan M., "Focuson: Nuclear Test Bans", *Journal of Peace and Research*, Vol. 24, No.2, Junuary pp. 105-10.

Djalal, Hasjim, "Potential Conflict in South China Sea: In Search of Co-operation", *The Indonesian Quarterly*, Vol. 18, No.2, Second Quarter 1990, pp. 127-132.

————, "The Relevance of the Concept of Joint Development to Maritime Disputes in the South China Sea," *The Indonesian Quarterly*, Vol.XXVII/1999, No.3, Third Quarter.

Djiwandono, Soedjati, "Neutralization: A New Hope of Southeast Asia", *Indonesian Quarterly*, Vol. 1, No. 2, January 1973, pp. 60-71.

Dorance John C., "United States Security Interests in the Pacific Islands", *World Review*, Vol. 23, No.1, April 84, pp. 4-20.

Dorrance, John, "Pacific Islands and US Security Interest: A New Era Poses New Challenges", *Asian Survey*, Vol. 29, No. 7, July 1998, p. 704.

Doty, Paul, "Nuclear Test Ban", *Foreign Affairs*, Vol. 65, No.4, Spring 1987, pp. 750-69.

Duncanson, Dennis, "Strategic Tension in Southeast Asia", *Conflict Studies*, Vol. 176, 1985, pp. 3-25.

Drifte, Reinhard, "Japan's Security Policy and Southeast Asia," *Contemporary Southeast Asia*, Vol. 12, No.3, December 1990, pp. 86-97.

Dunn, Lewis., "Containing Nuclear Proliferation", *Adelphi Papers*, No. 263, winter 91, pp. 1-75.

————————, "Nuclear Proliferation and World Politics", *The Annals of the American Academy of Political and Social Science*, Vol. 430, March 1997.

Ehibuya Eric., "Roaring Mice against the Tide: The South Pacific Islands and Agenda Building on Global Warming", *Pacific Affairs*, Vol. 69, No. 4, Winter 1996-97, pp.541-555.

Epstein, William., "Nuclear Testing threat", *Bulletin of the Atomic Scientists*, Vol. 46, No.6, July-August 1990, pp. 34-37.

Etcheson, Raig, "Reflections on Kampuchea", *Third World Quarterly*, Vol. 1.9, No.3, July 1987, pp. 976-85.

Evans, Gareth, Australia in the South Pacific", *World Review*, Vol. 28, No.2, June 1989, pp. 4-12.

Fee, Lew Eng, "Arms Control in South East Asia: A review of the debate", *Contemporary Southeast Asia*, Vol. 10, No.3, December 1988, pp.221-41.

Fieldhouse, Richard, China to mixed signals on nuclear weapons *Bulletin of the Atomic Scientists*, Vol. 47, No.4, May 1991, pp.37-48.

Finfied, Russel H., "South East Asia as a Regional Concept", *South East Asian Jourinal of Social Science*, Vol. 11, No.2, 1983, pp. 1-14.

Fry, Greg, " The South Pacific Nuclear Free Zone: Signficance and Implications", *Bulletin of the Concerned Asian Scholars*, Vol. 18, No.2, April-June, 1986, pp. 61-71.

————————, "The South Pacific Nuclear Free Zone", *SIPRI 1986*, pp. 501-507.

Galichev and Vasyugov, "ASEAN's Approach to Creating a Zone of Peace in Southeast Asia: Illusions and Reality", *Far Eastern Affairs*, No. 2, 1985.

Ganesan, N., "ASEAN's Relations with Major External Powers", *Contemporary Southeast Asia*, Vol. 22, No. 2, August 2000.

Gibson, Donald E., "Role of the Establishment in the Anti-nuclear Movement", *Sociological Spectrum*, Vol. 10, No.3, 1990, pp. 321-40.

Goldblat, Jozef, "The Treaty of Rarotonga: A modest but significant contribution to the cause of nuclear non-proliferation", *Transnational Perspectives*, Vol. 13, No.2, 1987.

Goralaski, Vladislav, "Nuclear Free Zones and Nuclear Disarmament", *Peace and the Science*, No. 2-3, 1988, pp. 101-104.

Guoxing, "Current Security Issues of Southeast Asia", *Asian Survey*, Vol. 26, No.9, September 1986, pp. 973-90.

Hahn, Bradley, "Southeast Asia's Miniature naval arms race, *Strategic Digest*, Vol. 15, No.10 October 1985, pp. 1232-39.

Hamzah, B.A., "Jurisdiction Issues and the Conflicting Claims in the Spratleys", *The Indonesian Quarterly*, Vol. 18, No.2, Second Quarter 1990, pp. 133-153.

Harrison, Richard A., "Neutralization Plan for the Pacific: Roosevelt and the Anglo-American Co-operation, 1934-37", *Pacific Historial Review*, Vol. 57, No.1, February 88, pp. 47-72.

Hegarty, David, "South Pacific Security Issues: An Australian Perspective," *Working Paper No. 147* (Canberra: The Strategic and Defence Studies Centre), December 1987.

Henming, Angeles M., ASEAN Security Co-operation After the cold war; Problem and Prospects", *Indonesian Quarterly*, Vol. 20, No.3, 1992, pp. 286-97.

Herring, George C., "America and Vietnam: The Unending War", *Foreign Affairs*, Winter 1991-92, pp. 104-19.

Heruouet, Gerand, "Cambodian Conflict: The Difficulties of Intervention and Compromise", *International Journal,* Vol. 45, No.2, Spring 90, pp. 258-91.

Hilali, Z.A., "US Interest and Policies in Southeast Asia", *Pakistan Horizon*, Vol. 44, No.3, July 91, pp. 67-96.

Hood, Steven J., "Beijing's Cambodia, Gamble and Prospects for Peace in Indo-China: The Khamer Rouge or Shihanouk?" *Asian Survey*, Vol. 30, No. 10, October 1990, pp. 977-91.

Hoong, Khong Kim and Abdullah (Abdul Razak), "Security Co-opertion in ASEAN", *Contemporary Southeast Asia*, Vol.9, No.2, September 1987, pp. 129-139.

————, "Japan and the Problem of Kampuchea", *Contemporary Southeast Asia*, Vol.2, No.3, December 1989, pp. 313-31.

Horn, Robert C., "Evolving Relationship among the Super Powers: The Implications for Southeast Asia", *Indonesian Quarterly*, Vol. 17, No.2, 1989, pp. 126-39.

————, "The Soviet Perspective", *Southeast Asia under New Balance of Power* (New York; Praeger, 1974).

Hussain, Saiyid Mubarak, "Nuclear-Weapon Free Zone: A Perception", *Pakistan Horizon*, vo, 43, No.1, 1990, pp.65-78.

Huxlay, Tim, "ASEANS Prospective Security Role: Moving Beyond the Indo-China Fixations", *Contemporary Southeast Asia*, Vol. 9, No.3, December 1987, pp. 194-207.

"Indian Ocean as Peace Zone" *Indian Background*, Vol. 2, No. 34, November 82, pp. 905-11.

Wanandi, Jusuf, "Regionalism in the Asia Pacific Region," *The Indonesian Quarterly*, Vol.XXVIII/, No. 2, Second Quarter, 2000.

Kapur, Ashok, "Dump the Treaty",*Bulletin of the Atomic Scientists*, Vol. 16, No.6, July-August 1990, pp. 21-22.

Karp Aaron, "Military Procurement and Regional Security in Southeast Asia", *Contemporary Southeast Asia*, Vol. 11, No.4, March 1990, pp. 334-62.

Kaur, Gurmeet, "Pakistan-Indonesia Bilateral Relations", *New Quests*, Vol.9, November December 1991, pp. 333-36.

Kelemen, Paul, "How Pol Pot came to Power", *Bulletin of Concerned Asian Scholars*, vol, 20, No.4, October December 1988, pp. 62-71.

Kende, Istvan, "History of Peace: Concept and Organisations from the late Middle Ages to 1870s" *Journal of Peace Research*, Vol. 26, No.3, August 1989, pp. 233-48.

Khamchoo, Chaiwat, "Japan's role in Southeast Asian Security: Plus a Change",*Pacific Affairs*, Vol. 64, No.1, spring 1991, pp. 7-22.

Khosla, Inder, "Indian Ocean as a Zone of peace",*International Studies*, Vol.2, No.4, October December 1982, pp. 417-44.

King, Warren., "South Pacific Security the Future of the Microstates", *World Review*, Vol. 29, No.4, December 1990, pp. 3-26.

Kittel, Gabriele and Others, "Nuclear Weapon-Free Zones: Re-evaluating the Political Concept of Peace Making",*Bulletin Peace Proposals*, Vol. 22, No.2, June 1991, pp. 217-26.

Kleck, Gray, and Mcelrath (Karen), "Effects of Weaponray on Human Violence",*Social Forces*, Vol.69, No.3, March 1991, pp. 669-92.

Klick, Donna J., "A Balkan Nuclear Weapons Free Zone: Viability of the Regime and Implication for Crisis Management", *Journal of Peace Research*, Vol. 24, No.24, June 1987, pp. 111-124.

Kortunov, Sergei, "START II and Beyond", *Bulletin of the Atomic Scientists*, Vol. 46, No.8, October, 1990, pp. 21-24.

Kumar, Chandra, "Indian Oceanic of Crisis or Zone of Peace?" *International Affairs*, Vol. 60, No.2, Spring 1985, pp.246-67.

Kuroyamai, Yoneji, "Kampuchaean conflict and ASEAN: A View From the Final State", *Japan Review of International Affairs*, Vol.3, No.1, spring/summer 1989, pp. 57-81.

Kusuma-Atmadja, Mochtar, "Some thoughts on ASEAN Security Co-operation: An Indonesian Perspective", *Contemporary Southeast Asia*, Vol. 12, No.3, December 1990, pp. 70-90.

Lange, David, "A Comprehensive test ban treaty", *New Perspectives*, Vol.6, 1987, pp. 11-13.

Leifer, Michael, "Regional Order in Southeast Asia", *The Round Table* No.255, July 1974, pp. 309-17.

Leng, Lee Yonk, " Access to Southeast Asian Water by Naval Powers: Some Problems and Ambiguities", *Contemporary Southeast Asia*, Vol. 9, No.3, 1987, pp. 208-20.

Luhulima, C.P.F., "ASEAN, the South Pacific Forum and the Changing Strategic Environment", *Indonesian Quarterly*, Vol. 20, No.2, 1992, pp. 207-18.

Lumpe, Lora, and others, "Third world missiles fall short", *Bulletin of the Atomic Scientists*, Vol. 48, No.2, March 1992, pp. 30-37.

Luthulima, C.P.F., "ASEAN-European Community Relations: Some Dimensions of International Co-operation", *Indonesian Quarterly*, Vol. 20, No.3, 1992, pp. 309-24.

Mackay (Don J.), "Some Legal Aspects of the South Pacific Nuclear Free Zone Treaty", *Regional Disarmament Workshop for Asia-Pacific*, New York: UN, 1991.

Malik, J. Mohan, "Nuclear Free Southeast Asia: The China Factor", *Current Affairs Bulletin*, Vol.65, No.10, March 1989, pp. 19-24.

————, "China and South Asian Nuclear Free Zone", *China Reports*, Vol. 25, No.2, April June 1980, pp. 113-17.

Mark, G. Rolles, "ASEAN: Where from or Where to?" *Contemporary Southeast Asia*, Vol. 13, December 1991.

Marshall, Russel, "New Zealand in the South Pacific", *World Review*, Vol.28, No.2, June 1989, pp. 13-32.

McGraw David, "From Kirk to Muldoon: Challenges and Continuity in New Zealand's Foreign Policy Priorities", *Pacific Affairs*,Vol.55, No.4, Winter 1982-83, pp, 640-659.

————, "New Zealand's Foreign Policy under National and Labour Governments:Variations on the 'Small State' Theme ?" *Pacific Affairs*, Vol. 55, No. 4, Winter1992-93, pp. 7-25.

Muetzenich, Rolf, "Nuclear Weapon-Free Zone in Southeast Asia", *Aussen Politik*, Vol. 48, No. 4, 1997, pp 391-401.

Mediansky, F.A., "Nuclear Free Security in the South Pacific?" *Australian Outlook*, Vol. 39, No.2, 1985, pp. 77-83.

Mehta, Usha, "Gandhian Movement against Nuclear Power", *Gandhi Marg*, Vol.2, No.1, April- June 1989, pp. 76-9.

Methuen, Phillip, "Deference to De Gualle: The French Approach to Security in the South Pacific",*Contemporary Southeast Asia*, Vol. 10, No.4, March 1989, pp. 385-410.

Millikan, Robert, "Half Life Deadly Show",*Bulletin of Concerned Asian Scholars*, Vol. 18, No.2, April-June 1986, pp. 90-91.

Misra, K.P., "Indian Ocean an a Zone of Peace: The Concept and Alternatives", *India Quarterly*, Vol. 33, No.1, January March 1977, pp. 19-32.

Mogami, Toshiki, "The South Pacific Nuclear Free Zone: A Fettered Leap Forward",*Journal of Peace Research*, Vol. 25, No.4, 1988.

Mohan, C. Raja, "The Treaty of Rarotonga",*Strategic Analysis*, October 1985, pp. 728.

______________, "The Treaty of Rarotonga", *Strategic Analysis*, October 1985, pp. 728.

Morrison, Charles, "US Security Relations with Southeast Asia: Possibilities and Prospects for the Clinton Administration", *Australian Journal of International Affairs*, Vol. 47, No.2, October 1993, pp. 239-262.

Morton, Robert, "Culture and Identity in the South Pacific: A Comparative Analysis", *Man*, Vol. 28, No.4, December 93, pp. 741-68.

Multan (Wojeiech), "The Past and Future of Nuclear-Weapon-Free Zones", *Bulletin of Peace Proposals*, Vol. 16, No.4, 1985, pp. 375-385.

Alagappa, Muthiah, "Regional Arrangements and International Security in Southeast Asia: Going Beyond ZOPFAN", *Contemporary Southeast Asia*, Vol. 12, No.4, March 1991.

Naidu, G.V.C., "Neutralization of South East Asia: Zone of Peace Freedom and Neutrality", *Strategic Analysis*, Vol. 10, No.3, June 1986, pp. 357-68.

______________, " ASEAN and Indo-China", *Strategic Analysis*, Vol. 15, No.12, March 1993. pp. 119-1204.

______________, "The Kampuchean Problem: Retrospect and Prospect",*Strategic Analysis*, 1987.

______________, "Japan and Southeast Asia", *Strategic Analysis*, Vol. 17, No.1, April 1994, pp. 109-22.

______________, "Post-Cold War Security Situation in Southeast Asia",*Strategic Analysis*, Vol. 14, No.4, July 1991, pp. 487-97.

______________, " China, Vietnam and the Kampuchean Problem", *Strategic Analysis*, Vol. 19, No.2, March 1986, pp. 202-16.

Namboodiri, P.K.S., "Essence of the Peace Zone Concept", *Strategic Analysis*, Vol. 31, No.2, March 1980, pp. 467-70.

Nathan, K.S., "Current US Policy in Asia: Principles, Problems and Performance", *The Indonesian Quarterly*, Vol.18, No.4, Fourth Quarter, 1990, pp. 320-331.

Narine, Shaun, "Asean and the Management of Regional Security", *Pacific Affairs*, Vol.71, No.2, Summer 1998.

Nelson (Kelly S.), "US-Vietnamese Normalization", *Asian Affairs*, vol 19, No.1, Springs 1993.

"Non Proliferation Treaty's 1990 Review", *Bulletin of the Atomic Scientists*, Vol. 46, No.6, pp. 2-48.

"Nuclear Weapon Free Zones, Europe needs a Churchill of Peace", *Problems of Peace and Socialism*, Vol. 16, No. 11, November 1988, pp. 43-51.

"Nuclear Weapon Safety", *Bulletin of Atomic Scientists*, Vol. 47, No.3, April 1991, pp. 29-40 (series of articles).

Ogunbanwo, Sola, "Introductory Statement on Nuclear Weapon Free Zones of Peace", *Regional Disarmament Workshop for Asia and Pacific*, (New York: United Nations 1991), pp. 155-157.

Pande, Savita, "Tlatelolco Treaty: How Successful", *Strategic Analysis* Vol.22,No.1, April 1998, pp 35-48.

______________ , "Prospects of Nuclear Weapon-Free Zones: South-East and Central Asia, *Strategic Analysis*, Vol.22, No.6, September 1998.

______________ , "Treaty of Rarotonga: A Nuclear Free Pacific", *Strategic Analysis*, Vol.22, No.2, May 1998, pp 195-207.

Pao-Min, Chang, "New Scramble for the south China Sea Islands", *Contemporary Southeast Asia*, Vol.12, No.1, June 1990, pp. 20-39.

"Pattern in Southeast Asia", *Foreign Affairs Malaysia*, Vol.4, No.4, December 1971, pp. 47-52.

Payne, Anthony J., "Fiji effect: A Review of Trends in South Pacific Politics", *Round Table*, Vol. 312, October 1989, pp. 440-46.

Piper, John., "Don't sit under the Coconut Tree with Anybody Else but Me: Australian Policy Towards the South Pacific", *Australian Journal for International Affairs*, Vol. 44, No.2, August 1990, pp. 119-30.

Power, Paul F., "The South Pacific Nuclear Free Zone", *Pacific Affairs*, Vol. 59, No.3, Fall 1986, pp. 457-75.

Prakash Shri., "Prospects of Cooperative Security and Peace in Asia-Pacific During Twenty-first Century", *Journal of Indian Ocean Studies*,Vol.9, No.1, April 2001.

Puddington, Arch, Pol Pot in Retrospet, *Commentary*, Vol. 83, No.4, April 1987, pp. 49-54.

Quinlan, Michael, "Nuclear Weapons an the abolition of War", *International Affairs* Vol. 67, No.2, April 1991, pp. 293-302.

Raman, Sudha, "Spratly Dispute: Creeping Invasion, Diplomatic Evasion", *Journal of Indian Ocean Studies*, Vol.9.No.-1 April 2001.

Rana, Swadesh, "A Permanent US Presence in the Indian Ocean", *Strategic Analysis,* Vol.3, No.2, May 1979, pp. 77-80.

Rittberger, Volker and others, "Nuclear Weapon-Free Zones and Strategies of Denuclearization: Instruments for Nuclear Disarmament, Security and Alternative Peace Policy?" *Law and State*, Vol. 35, 1987, pp. 40-62.

Roesnadi, Sutomoo, "ASEAN and the Great Powers", *Indonesian Quarterly*, Vol. 1, No.4, July 1973, pp. 15-25.

Roff, Sue Rabbitt, "New Caledonea: Decolonisation and Denuclearisation in the Pacific", *Third World Quarterly*, Vol.8, No.2, April 1986, pp. 621-38.

Ross Robert S., "China and Post-Cambodia Southeast Asia: Coping with success", *Annals of that American Academy of Political and Social Sciences*, Vol.519, January 1992, pp. 52-66.

——————, "China's Strategic View of Southeast Asia: A region in Transition", *Contemporary Southeast Asia*, Vol. 12, No.2, September 1990, pp. 101-19.

Ross, Andrew L., "Growth debt And Military Spending in Southeast Asia", *Contemporary Southeast Asia*, Vol. 11, No.4, March 1990, pp. 243-64.

Ross, Ken, "Asia and the Security of the South Pacific's Island States", *Survival*, Vol.38, No.3, Autumn 1996, pp.129-43.

Rudolf, Peter, "Non-Proliferation and International Export Control" *Aussen Politic*, *Quarter* Vol. 42, No.4, 1991, pp. 390-401.

Sadleir (David), "Rarotonga: In the Footsteps of Tlateloco", *Australian Foreign Affairs Record*, Vol. 58, No. 9, September/October, 1987.

Saiyad, Mubarak Hussian, "Nuclear Weapon Free Zone: A Perception", *Strategic Digest,* Vol. 20, No.12, December 1990, pp. 3311-17.

Scalapino, Robert A., "Basic trends in the Pacific Asian region in the 1980s", *Journal of Asiatic Studies*, Vol. 27, No.1, 1984, pp. 307-17.

Scarlott, Jennifer, "Nuclear proliferation after the cold war", *World Policy Journals*, Vol. 8, No.4, Fall 1991, pp. 687-710.

Schellhorn, Kai M., "Politics in the South Pacific Region", *Contemporary Southeast Asia*, Vol. 13, No.2, September 1991, pp. 188-99.

Schwarz, Wolfgong, "Washington Strategy in Pacific Oceans: Plan for a NATO Counterpart in Far East", *New Perspective*, Vol. 14, June 1984. pp. 4-8.

Shah, Prakash, "Nuclear Non-Proliferation Implications and the NPT Review: An Indian Perspective", *Strategic Analysis*, Vol. 16, No.2, May 1993, pp. 139-46.

Sheleping, M., "Nuclear Free Zones: A Means of reducing the war threat", *International Affairs*, Vol. 3, March 1985, pp. 15-20.

Shepherd, George W., "Developing Collective Self-reliance in the Zone of Peace", *IDSA Journal*, Vol. 15, No.1, July-September 1982, pp. 19-33.

Simon, Sheldon W., "United States and Conflict Reduction in Southeast Asia", *Contemporary Southeast Asia*, Vol. 12, No.2, September 1990, pp.83-100.

______________ , "United States interest in Southeast Asia: The Future Military Presence", *Asian Survey*, Vol. 31, No.7, July 1991, pp. 662-75.

______________ , "Two Southeast Asian and China: Security Perspectives" *Asian Survey*, Vol. 24, No.5, May 1985, pp. 519-33.

______________ , "The Nuclear Non-prliferation Regime after the NPT Review and Extension Coference", *SIPRI Yearbook 1996: Armaments, Disarmaments and International Security* 1996.

______________ , and Howlett (Darryl) "Nuclear Non-proliferation: The way forward", *Survival* Vol. 32, No.6, November/December 91, pp. 183-99.

______________ , "1990 Review Conference of the Nuclear Non-proliferation Treat: Pointer to the future diplomatic accident", *Round Table*, Vol.318, April 91, pp. 139-54.

Singh, Swaran, "Problems of Nuclear Weapon Free Zone", *Third Concept*, Vol. 3, No.33, November 1989, pp. 34-36.

Singh, Bilveer, "Soviet Military Presence in the Asia-Pacific Region: Implications for SEA", *Indonesian Quarterly*, Vol. 15, No.3, July 1987. pp. 405-18.

______________, "Nuclear Weapon Free Zone in Southeast Asia: The Initial Idea", *Asian Defence Journal*, No. 12, December 1989, pp. 38-41.

Singh, Vishal, "ASEAN and the Security of the Southeast Asia", *International Studies*, Vol. 23, No.3, July-September 1986, pp. 217-23.

Singh, Jasjit, "Future of NPT", *World Forms*, Vol. 13, No.11-12, November-December 1992, pp. 24-7.

______________ , "Indian oceanic Ozone of Peace or Conflict?", *Strategic Analysis*, Vol. 10, No.6, September 1986, pp. 649-63.

______________, "Developing Countries and Nuclear Issues", *Strategic Analysis*, Vol.6, 1987, pp. 11-13.

Sinuraya, Thomas, "Disperse the Clouds of Tension in South-East Asia", *World Marxist Review*, Vol. 29, No.4, April 1986, pp. 85-88.

Siracusa, Joseph M., "Nuclear Diplomacy, the Arms Race, and Arms Control: From Truman to Reagan",*Australian Journal of Politics and History*, Vol. 29, No.2, 1983, pp. 308-17.

Snider, Lewis W., "Political dimensions of Military spending and Debt Service in Southeast Asia", *Contemporary Southeast Asia*, Vol. 2, No.4, March 1990, pp. 278-305.

Sodhy, Pamela., "Survey of US Post-Vietnam Policy and the Kampuchean Dilemma, 1975-89: A Southeast Asian View", *Contemporary Southeast Asia*, Vol. 2, No.3, December 1989 pp. 283-312.

Soeastro, M. Hadi, "Japan and Southeast Asia: A Regional or Global Question", *Indonesian Quarterly*, Vol. 1, No.4, July 1973, pp. 27-35.

Soewondo, Purbo S., "Proposals relating to the Establishment of a South-East Asian Nuclear Weapon Free Zone", *Regional Disarmament Workshop for Asia Pacific*, (New York: UN, 1991).

Solidum, Estrella D., and Morales (Nalalia M), "Comparative Study of Collective Security Plans for Southeast Asia", *Asia Pacific Community*, Vol. 18, Fall 1982, pp. 26-41.

Solingen, Etel, "Asean, Quo Vadis ? Domestic Coalitions and Regional Cooperation", *Contemporary Southeast Asia*, Vol. 21, No.1, April 1999.

"Southeast Asia", *World Focus*, Vol. 8, No.1, January 1987, pp. 1-32.

Sridharan, Kripa, "India and Asean: The Long Road to Dialogue", *The Round Table*, October 1996, pp. 465-477.

Srivastava, B.K., "START Achievements and Limitations", *Link*, Vol. 34, No.1, 18 August 1991, pp. 45-48.

Stevenson, Charles H., "U.S. Foreign Policy in Southeast Asia: Implications for Current Regional Issues", *Contemporary Southeast Asia*, Vol. 4, No.2, September 1992, pp. 87-111.

Stubbs, Richard, "Reluctant Leader Expectant Followers: Japan and Southeast Asia", *International Journal*, Vol. 46, No.4, Autumn 1991, pp. 649-67.

Subrahmanyam, K., "Peace Zone Concept Dead", *World Focus*, Vol. 7, No.8, No.80, August 1986, pp. 3-6.

Sudo, Sueo, "From Fukuda to Thakeshita: A Decade of Japan ASEAN Relations", *Contemporary Southeast Asia*, Vol. 10, September 1988, pp. 119-143.

————————, "Japanese-ASEAN Relations: New Dimensions in Japanese Foreign Policy", *Asian Survey*, Vol. 28, May 1988, pp. 509-25.

Sukma, Rizal, "Jakarta Beijing relations and Security Challenges in Southeast Asia", *Indonesian Quarterly*, Vol. 18, No.4, 1990, pp. 80-86.

______________, "Security Arrangements in Southeast Asia: A Challenge for ASEAN in Post-Cold War Era", *Indonesian Quarterly*, Vol. 20, No.3, 1992, pp. 273-85.

______________ , "Asean and the Asean Regional Forum: Should The Driver Be Replaced?", *The Indonesian Quarterly*, Vol. XXVII, 1999, No. 3.

Simon, Sheldon W., "US interest in Southeast Asia: The Future Military Presence", *Asian Survey*, Vol. 31, No.7, July 1991, pp. 662-25.

Sundaraman, Shankari, "The Asean Regional Forum: Reassesing Multilateral Security in the Asia Pacific", *Strategic Analysis*, July 1998.

"Supply-side Controls on Nuclear Proliferation", *Orbis*, Vol. 36, No.2, Spring 1992, pp. 211-38.

Suryanarayanan V, "Sri Lanka and the Indian Ocean as a Zone of Peace", *Strategic Analysis*, Vol. 13, No.4, July 1990, pp. 453-62.

____________, and Jose, Alice, "Inter-Communist Conflicts: China Vietnam and Cambodia", *Link*, Vol. 34, No.47, 5 July 1992, pp.28-32.

_____________, "Settlement In Sight", *World Focus*, Vol. 9, No.6, June 1988. pp. 3-4.

Suter, Keith D., "US and the South Pacific: A Benign Neglect?" *Contemporary Review*, Vol. 258, No.1503, April 1991, pp. 184-98.

Sutopo, A.R., "US and Soviet Military Presence in Southeast Asia: An Overview", *Indonesian Quarterly*, Vol. 18, No.4, 1990, pp. 287-99.

Synder, Craig A., "Australia's Pursuit of Regional Security into the 21st Century", *Journal of Strategic Studies*, Vol. 21, No.4, December 1998, pp 1-17.

Syquia, Enrique P., "Regional Peace Through Peaceful Co-existence: The ASEAN experience", *Korean Journal of International Studies*, Vol. 20, No.4, Winter 1989, pp. 585-604.

Thakur, Ramesh, "Creation of the Nuclear-Free New Zealand Myth: Brinkmanship without a Brink", *Asian Survey*, Vol. 29, No. 10, October 1989, pp. 919-939.

______________ , "The Last Bang before a Total Ban: French Nuclear Testing in the Pacific", *International Journal* Vol. No. 1, Summer 1996, pp.466-486.

________________, "Nuclear Weapon Free South Pacific: A New Zealand Perspective", *Pacific Affairs*, Vol. 58, No.2, 1985.

________________, "A Dispute of Many Colours: France, New Zealand and the Rainbow Warrier Affairs", *World Today*, Vol. 42, No.12, December 1986.

Thayer, Carlyle A., "Prospects for peace in Kampuchea: Soviet Initiatives and Indo-Chinese Responses", *Indonesia Quarterly*, Vol. 17, No.2, 1989, pp. 157-72.

Tinger, E. Jakub, "Nuclear Free Zones and Zones of Peace: The Position of USSR, "*Development and Peace*, Vol. 8, No.2, Autumn 1987, pp. 70-78.

To Lee Lai, "Managing Potential Conflict in the South China Sea: Political and Security Issues", *The Indonesian Quarterly*, Vol. 18, No.2, Second Quarter, pp. 154-163.

Toba, Reifiro, "Japan's Southeast Asian Policy in the Last Decade", *Asia Pacific Community*, No. 15, Winter 1982, pp. 30-43.

UM, Khalharya, "Thailand and the Dyanmics of Economic and Security Complex in Mainland Southeast Asia", *Contemporary Southeast Asia,* Vol. 13, No.3, December 1991, pp. 245-70.

"United Nations Initiatives in making Indian Ocean a Zone of Peace", *New Perspective*, Vol. 10, No. 1, 1980, pp. 34-5.

Vashisht, Dashrath Kumar, "Australia and French Nuclear Tests", *Strategic Analysis,* vol 8, No.7, October 1984, pp. 668-76.

Vatikiotis, Michael, "Zone of Discord", *Far Eastern Economic Review*, 14 January 1986.

Velencia, Mark J. and Marsh James Barney, "Access to Straits and Sea lanes in South East Asian Seas: Legal, Economic and Strategic Considerations", *Journal of Maritime Law and Commerce,* Vol. 16, No. 64, October 1985, pp. 414-22.

VU Tung, Nguyer, "Vietnam ASEAN Co-operation in Southeast Asia", *Security Dialogue*, Vol. 24 No. 1, March 1993, pp. 85-92.

Wagher, R. Harrison, "Nuclear Deterrence Counter Force Strategic and the Incentives to Strikes First", *American Political Science Review*, Vol. 85, No.3, September 1991, pp. 727-50.

Wanandi, Jusuf, "The USSR in Asia-Pacific: an ASEAN view", *The Indonesian Quarterly*, Vol. 18, No.4, Fourth Quarter, 1990 pp. 300-311.

____________, "Southeast Asian Security in Perspective", *Indonesian Quarterly*, Vol. 7, No.4, October 1979.

__________, "Peace and Security in Southeast Asia", *Indonesian Quarterly*, Vol. 19, No.4, 1991, pp. 313-25.

Williams, Michael C., "New Soviet Policy Towards Southeast Asia: Re-orientation and Change", *Asian Survey*, Vol. 31, No.4, April 1991, pp. 364-77.

Woltsthal, Jan Brook, "Nuclear-Weafor-Free Zones: Coming of Age," *Arms Control Today*, Vol. 23, No.2, March 1993.

Yost, David S., "Nuclear Debates in France", Survivals, Vol. 36, No.4, Winter 1994-95, pp. 113-34.

Index